MECHANIZATION OF REASONING IN A HISTORICAL PERSPECTIVE

POZNAŃ STUDIES
IN THE PHILOSOPHY OF THE SCIENCES AND THE HUMANITIES

VOLUME 43

The address: **prof. L. Nowak, Cybulskiego 13, 60-247 Poznań, Poland.**
FAX: (061) 477-079
E-mail: epistemo at PLPUAM11.AMU.EDU.PL

Witold Marciszewski
and
Roman Murawski

MECHANIZATION OF REASONING IN A HISTORICAL PERSPECTIVE

Amsterdam - Atlanta, GA 1995

∞ The paper on which this book is printed meets the requirements of "ISO 9706:1994, Information and documentation - Paper for documents - Requirements for permanence".

ISSN 0303-8157
ISBN 90-5183-790-9 (Bound)

Printed in The Netherlands

To the Great Masters of the Formalization of Reasoning

Stanisław Jaśkowski

Stanisław Leśniewski

Jan Łukasiewicz

Alfred Tarski

in the Centenary of the Lvov-Warsaw School (1895–1995)

ACKNOWLEDGEMENTS

This volume is written jointly by Witold Marciszewski, who contributed the introductory and the three subsequent chapters, and Roman Murawski who is the author of the next ones – those concerned with the 19th century and the modern inquiries into formalization, algebraization and mechanization of reasonings. Besides the authors there are other persons, as well as institutions, to whom the book owes its coming into being.

The study which resulted in this volume was carried out in the Historical Section of the research project *Logical Systems and Algorithms for Automatic Testing of Reasoning*, 1986-1990, in which participated nine Polish universities; the project was coordinated by the Department of Logic, Methodology and Philosophy of Science of the Białystok Branch of the University of Warsaw, and supported by the Ministry of Education (some of its results are reported in (Srzednicki (Ed.) 1987). The major part of the project was focussed on the software for computer-aided theorem proving called Mizar MSE (Multi-Sorted first-order logic with Equality, reported in (Marciszewski 1994a)) due to Dr. Andrzej Trybulec. He and other collegaues deserve a grateful mention for a hands-on experience and theoretical stimulants owed to their collaboration.

The authors are grateful to Professor Ewa Żarnecka-Biały as the main referee of the contributions produced in the Historical Section, and to Professor Olgierd Wojtasiewicz who helped to translate Chapters 2, 3 and 4 into English. Also the organizational assistance of the University Department for Research Administration, run by Mrs. Bożena Sudnik, is sincerely appreciated.

As for the book itself, the authors are indebted to Professor Leszek Nowak for his encouragement, patience, and editorial suggestions. The typesetting and ingenious technical assistance is due to Mr. Jarosław Sokołowski, while Mrs. Halina Jastrzębska Marciszewska put the finishing touches to the volume through proofreading, the compiling of indexes and references, and typographical refinements.

The authors hope to enjoy comments from benevolently critical Readers. For the purposes of the so hoped discusion, here are the authors' correspondence addresses.

WITOLD MARCISZEWSKI
Professor (ordinary) at the University of Warsaw, Department of Logic, Methodology and Philosophy of Science (Head) in the UW Białystok Branch. Liniarskiego 4 (street), 15-420 Białystok, Poland. E-mail: "witmar@plearn.edu.pl".

ROMAN MURAWSKI
Professor at Adam Mickiewicz University in Poznań, Faculty of Mathematics and Computer Science, Department of Mathematical Logic. Matejki 48/49 (street), 60-769 Poznań, Poland. E-mail: "rmur@pluam11.amu.edu.pl".

Poznań and Białystok, December 1994

Table of Contents

CHAPTER ONE

FROM THE MECHANIZATION OF REASONING TO A STUDY OF HUMAN INTELLIGENCE

There are no bare facts in history, be it political history, be it a story of ideas. Historical narration tells about past facts, but there are as many stories as are *perspectives* into which we put data found in sources. The choice of a perspective may depend on axiological assumptions, as well as the position in time assumed by a historian. The latter obviously divides into (i) the standpoint of the past under study (e.g., an attempt to read Aristotle in the light of Aristotle), (ii) the standpoint of the present state of affairs (as, e.g., in Łukasiewicz 1951), and (iii) the standpoint of an envisaged future development.

It was Jan Łukasiewicz who was the pioneer of approach (ii), as expressed in the title of his *Aristotle's Syllogistic from the Standpoint of Modern Formal Logic*, 1951. Among his seminal achievements in this field there was an interpretation of Stoic logical writings in the light of modern propositional logic.

Approach (iii) highlights those facts which one deems relevant to a forecasted course of events. The forecast to be substantiated in this chapter, and to shed light at the content of this volume, is as follows. There starts a process of merging logic with *cognitive science*, the latter being the theory of perception and intelligence advanced in terms of information-processing; this is not to mean any loss of autonomy on either side, rather the emergence of a new research area to which both sides essentially contribute. When seen in this perspective, the development of logic leads in the direction which was hardly expected by the founding fathers. In their intention, logic should have been an indispensable tool of research. In fact, though, discoveries do not happen to result from a conscious use of logical inference rules. Instead, logic made it possible to create artificial reasoners, and these prove apt models to help us in a better understanding of natural reasoning (even if they function as negative models to yield a contrastive background); and once we see reasoning as a method of problem-solving, we enter on a study of intelligence.

This volume is to show how the main stream of development of logic has led to that result. The idea of formalization of reasoning had originated in the Middle Ages, and reached its maturity in modern logic, that is, in Frege (1879, 1893, etc.), Peano (1891) and Whitehead and Russell (1910-1913); the most sophisticated stage of formalization is due to Hilbert (1928, 1934, 1939), Łukasiewicz (1925, 1930, 1951), Gödel (1930, 1931, etc.), Tarski (1931, 1935, 1956), Jaśkowski (1934), Leśniewski (1992 – collected works), etc.

Within that formalization trend there was the result of the utmost importance for the mechanization of reasoning, namely the algebraization of logic in the form of binary algebra (claimed by Leibniz and effected by Boole and others), the latter having been combined with binary arithmetical notation and with the two-states functioning of electrical circuits; to that came methods of arithmetizing syntax and, moreover, methods of reducing the whole of logic to the binary algebra (the elimination of quantifiers). That development resulted in the present state of logic and essentially contributed to the rise of computers. Those, in turn, made it possible to mechanize deductive reasoning which proves a suitable base to start with an inquiry into the nature of intelligence.

In such a historical perspective the development of logic is viewed from the position taken in this volume. This is the standpoint of the present relations between logic and cognitive science, the present being seen as pregnant with an expected future development – according to Leibniz's dictum *praesens gravidum est futuro.*

1.1. Von Neumann's project related to cognitive science

1.1.1. Is logic a device-independent system, or is it relative to a device (e.g., a brain) carrying out logical operations? The import of this question seems to escape attention of quite a number of logicians, as well as historians and philosophers of logic, who feel proud for overcoming the 19th century *psychologism*, attacked by Husserl, Frege, Łukasiewicz, and other outstanding authors.

Psychologism construes the laws of logic as ruling mental operations, that is, a behaviour of a reasoning subject; when the subject in question is considered like a tool of reasoning, it can be called a device. A new, even more radical, version of antipsychologism is due to Karl Popper (1967) with his programme of 'epistemology without a knowing subject'. Fortunately, there is no dilemma of whether such a philosophy (supported by some good reasons, indeed) should be maintained, or superseded by the new approach to relativize logic to a reasoning device. The new one is to be seen as a complementing alternative, which does not contradict the views advanced by the former classics.

However, there happen to be authors who do not acknowledge that alternative. This is why this chapter starts from recalling John von Neumann's project which Helmut Schnelle (1988) calls naturalization of logic, and which may be also called PHYSICALIZATION OF LOGIC.[1] These are fitting terms, for the project consists in considering mental mechanisms as the basis of logic and mathematics, provided that those described as mental are somehow reduced to neural mechanisms; the latter certainly are devices belonging to the natural, or physical, world. Here is a radical statement of that programme (von Neumann 1951/1963, p. 311).[2]

> There is an equivalence between logical principles and their embodiment in a neural network, and while in the simpler cases the principles might furnish a simplified expression of the network, it is quite possible that in cases of extreme complexity the reverse is true. All of this does not alter my belief that a new, essentially logical, theory is called for in order to understand high-complication automata and, in particular, the central nervous system. It may be, however, that in this process logic will have to undergo a pseudomorphosis to neurology to a much greater extent than the reverse.

Since the central nervous system is treated as a mechanism and – in a certain respect – classed with computers, the programme of nat-

[1] In this chapter, unlike in the following ones, small capital letters (not only italics) are used to distinguish the key concepts. Those distinguished by capitals are meant as most important for the theoretical approach taken in this chapter.

[2] The quotation after (Schnelle 1988, p. 541). The phrase 'physicalization of logic' to stand for von Neumann's approach is found in (Marciszewski 1994, p. 172). The term 'naturalization' is quite fitting as it comprises both physical and biological aspects but the term 'natural' appears in other logical contexts (e.g., natural deduction), and so it may lead to misunderstandings; hence the adjective 'physical' is preferred in this chapter.

uralizing (physicalizing) logic can equivalently be called a *machine-oriented approach to logic.* The latter phrase is self-explaining while the former is more convenient as giving rise to the adjective 'naturalized' (or 'physicalized'), hence both will be employed. Moreover, both are related to the concept of cognitive science which we shall need to make our discussion more general. What should we understand under 'cognitive science'?

1.1.2. While the traditional epistemology deals with an abstract subject of cognition which endeavours to find a truth, the modern COGNITIVE SCIENCE sees a subject as a device which can be called a thinking machine. It is construed as follows.

A THINKING MACHINE is an information-processing system that has a hardware component, the processing being aimed at the growth of information; INFORMATION-PROCESSING is found in a continuous interplay with DATA-PROCESSING, the former being construed as operating on abstract objects (numbers, propositions, etc.), the latter as operating on physical tokens (e.g., numerals produced with ink, or by magnetization of some spots, etc.) which represent pieces of information. Let such a machine be also called a MIND, for short, this term covering both natural and artificial minds.[3]

LOGIC is the theory which deals with a special kind of information-processing, namely that which preserves the truth of information. Thus it can be seen as a preliminary part of cognitive science, provided it takes into account thinking machines, i.e., devices capable of processing information. Among such machines there is the central nervous system as discussed by von Neumann. Let that branch of logic, as dealing with minds, be termed *mind-theoretical logic.*

There is an opportune path from mind-theoretical logic to cognitive science, the path that leads through the theory and practice of mechanized deduction. The term *deduction* is used deliberately

[3]Thus, the definition is suited for animal, human, and artificial minds. It does not cover bodyless minds (if there are any), and those whose knowledge is not extendable; this is due to the intention to refer to so limited a class alone for purposes of the present discussion (no 'materialistic' tenet is hidden beyond this terminological option, as treating minds in terms of machines can be reconciled with a non-materialistic point, e.g., along the lines suggested by Leibniz).

to distinguish deductive reasoning, as defined by resorting to predicate logic, from other inferences, as probabilistic, statistical, etc.[4] In what follows (in this chapter and the whole volume) the term REASONING is meant in that narrowed sense, that is, to stand for deductive reasoning, or deduction, for short. Let us trace that path. To start with, two meanings (at least) of the term 'mechanization' should be distinguished for our purposes.

(i) In a broader sense, 'mechanization' is synonymous with 'formalization', the latter meaning the existence of a mechanical, i.e., algorithmical procedure to check validity of a proof; that use of the adjective 'mechanical' is usual in contexts dealing with FORMALIZATION. For instance, Davis (1988b, p. 317) writes as follows (about Post's contributions to computer science).

> Post studied the problem of finding algorithms by which it could be mechanically determined whether particular formulas in the language of *Principia Mathematica* could be derived using the rules of the language. For this purpose Post replaced the dubious "primitive propositions" [see below] with the purely syntactic principle of *modus ponens*:
>
> "$\vdash P$" *and* "$\vdash P \supset Q$" *produce* "$\vdash Q$"
>
> (which of course is what Russell and Whitehead actually used). From this point of view, the interesting thing about the logicist program, is that it implied the success of finding such algorithms would have led to the mechanization of large part of mathematics. Post solved only the first part of the problem; he found algorithms for the part of *Principia Mathematica* that we now call propositional calculus. Post's efforts to extend these results led him to consider formal operations on strings in the most general context, which he called *productions*. [...] Post's productions are ubiquitous in computer science.

The purely syntactic principle of *modus ponens* as formulated by Post is by Davis opposed to its informal statement in *Principia*

[4]Thus defined deduction is roughly identical with the method used for proving mathematical theorems. Only 'roughly', because mathematical induction, though being a deductive (in spite of its name) method of reasoning, cannot be justified by any law of predicate logic. Problems of mechanization of mathematical deduction form a special field of research; see, e.g., (Basin 1994) and (Moore 1994).

which reads: *A proposition implied by a true premise is true*; this is not a form manageable by a mechanical device, while Post's is – owing to the fact that it desribes just purely syntactic relations, i.e. those occurring among symbols as physical tokens.

(ii) In the narrower sense, 'mechanization' denotes a state in which formalization is suited for a definite device. This means that there exists such a device and a software to operate on it in order to process sentences (syntactically defined strings of tokens) according to an algorithm involved in the given formalization. If the device in question is a computer, then one may speak of *computerization of reasoning.* This concept of MECHANIZATION is found in an early contribution by Hao Wang (1970, p. 225 ff.) entitled 'Toward mechanical mathematics'. An illuminating context is provided by the abstract (preceding the paper) in which we read.

> [...] It is suggested that time is ripe for a new branch of applied logic [...]. This discipline, it is believed, will in the not too remote future lead to proofs of difficult new theorems by machine. An easier preparatory task is to use machines to formalize proofs of known theorems. [i.e., to present them as formalized proofs, according to logical rules presupposed in the program - W.M.] This line of work may also, it is thought, lead to mechanical checks of new mathematical results comparable to the debugging of a program.

The results reported in this paper are those of his author himself and of other researchers; among the latter are some who used johniac and succeeded in proving 38 theorems of *Principia Mathematica* (e.g., the proof of *2.45, found in 12 minutes), while in 14 cases the computer failed to find the proof because of time or space limitations. This sounds encouraging, if one compares the power of modern computers with that of johniac used in the 1950s; Davis (1988a, p. 171) recalls that johniac computers had a total operational memory of 41000 bits).

To sum up, formalization, or mechanization in sense (i), consists in using an ABSTRACT MACHINE which is conceived as a system of 'rules of thumb' — a term used by Alan Turing (1947, p. 107). Such rules control the process which is carried out simply by following a list of unambiguous instructions referring to finite discrete configurations of whatever kind. Mechanization in sense (ii) comes only then when there is physical device and a mapping between this device and an abstract machine. Thus, to find such a mapping was the fundamental engineering problem which must have been solved to pass from abstract machines (i.e., formalized systems) having been built by Turing, Hilbert, Gentzen, and others, to what we nowadays call computers.

1.1.3. The decisive step has been made by Claude Shannon (1938) in his master's thesis. It consists in interpreting *true* and *false* as corresponding to a switch being closed or open, respectively. The logical operators 'and' and 'or' then correspond to switches being connected in series and parallel, respectively. Thus the connection has been established between propositional logic and binary digital circuitry. This having been done, the methods of 'reducing' predicate logic to propositional logic, i.e. of eliminating quantifiers, due to Skolem, Hilbert, and others (extensively discussed in this volume) enabled the use of binary digital circuitry for mechanizing proofs in predicate logic.

One more step should have been made to pass from abstract mechanisms to practical computerization of reasoning. This one is due to Herbrand's and Gentzen's cut-free formalisms of the predicate calculus (in this book treated with due attention). The CUT is a schema representing a number of inference rules, one of them being the familiar *ponendo ponens*, whose use in a proof requires some invention from the reasoner (premisses from which the conclusion is being 'cut' must be found). On the other hand, in a proof produced with the cut-free formalism, each step is determined by the syntactic structure of the formula processed, hence the need of invention is extremely reduced, and the whole procedure becomes fairly mechanical. Gentzen demonstrated that any proof involving the cut rule can be transformed into a proof in which the cut does not appear, and this implies availability of a mechanical procedure for any proof formalized in predicate logic.

All those discoveries and techniques pave the way to realizing how an abstract system of information-processing rules can be implanted into a physical machine, so that those rules function like natural laws. Now, imagine that there are in Nature systems which are controlled by such physical laws corresponding to logical rules; the multitude of such laws proves much more comprehensive than that of those treated in theoretical logic. Hence the discovery of those laws should provide us with a new and much ampler logical system, being at the same time a system of natural regularities. There are such systems, indeed, namely animal brains. Reconstructing logic from the laws of their functioning would amount to an accomplishment of von Neumann's project of physicalization, or naturalization, of logic.

However, even if not accomplished as yet, this project is apt to function as a research programme for cognitive science. In the light of the MECHANIZATION OF REASONING logical laws reveal their new aspect as ways of behaviour of machines, hence as a factor belonging to the physical world. Once having been acknowledged in that function, they are supposed to govern certain activities of organic machines too; the empirical evidence gained so far does confirm this assumption.

This gives rise to new tasks: one should go far beyond the crude laws of Boolean algebra and seek for other, subtler, mechanisms of reasoning as well, and one should go beyond that specific kind of information-processing which amounts to logical transformations; from the cognitive point of view, there are more laws of information-processing than those treated by logic. However, in this chapter we confine ourselves to the logical part of cognitive science which deserves to be called *mind-philosophical logic.* Here is a comment on this terminological stipulation.[5]

There is a piece of philosophy, called philosophy of mind, and there is a kind of logic, called philosophical. PHILOSOPHY OF MIND is the host of issues centred around the mind-body problem, i.e., the problem of relations between these two categories of entities. To the problems of this couple Karl Popper added the issue of relations between each of them and the realm of information, called by him 'the third world', which includes e.g., *statements in themselves* (as once postulated by Franz Brentano; cf. Popper 1967, Popper 1982). Data-processing belongs to the world of physical bodies, while information-processing belongs to the third world, and there is the problem of how the mind (the second world) is related to either of them. These questions prove much relevant to some logical issues.

---

[5] As far as the authors know, the term 'mind-philosophical logic' was first used in (Marciszewski 1994, p. 15 ff.) where a more extensive motivation and a research programme of the so designed discipline is suggested.

PHILOSOPHICAL LOGIC is that which either takes advantage of philosophical concepts and assumptions or helps philosophy in clarifying its concepts and justifying its statements. These two tasks happen to be so different from each other that we would need a designation to suggest that looseness of connections, e.g., 'philosophy-oriented logic' (the orientation being in either direction, as indicated above). However, when bearing in mind this proviso, we can use the more handy term 'philosophical logic'.

The mentioned features should be included in the meaning of the phrase 'mind-philosophical logic'. For instance, there are logical formalisms to describe computer programs which, at the same time, can clarify some concepts referring to mental behaviour in our copying with tasks. Such is *dynamic logic* – dealing both with formulas and with programs – as well as its variety called *algorithmic logic*; it can be applied, e.g., to deontic concepts, or to explain the modal concept of possibility as menageability by a program (cf. Oberschelp 1992, Harel 1984).

However, in this discussion we shall take into account the other relation between logic and philosophy – that which consists in providing logic by philosophy with what Popper (1982) called a *metaphysical research programme*. Which mind-philosophical issues are here at stake will be seen in the sequel of this Chapter.

1.2. The Leibniz-style Cybernetic Universe

1.2.1. A substantial concept to give foundations to the following discussion is that of cybernetic universe. Let us start from explaining why we need the terms 'universe' and 'cybernetic'.

The idea of CYBERNETIC UNIVERSE goes back to Leibniz, though he did not use this modern designation. Instead, he spoke of *machines*, or *automata*, in so a general sense that he mentioned 'mental machines' (*machine spirituales*) to refer to algorithmic linguistic devices (cf. his *Accessio ad arithmeticam infinitorum*), and 'incorporeal automata', or 'natural automata', or else 'divine machines' to refer to minds and organisms (cf. *Monadology*, par. 18 and 64). The term 'incorporeal' does not contradict the definition of mind stated above; for, in the context in question, Leibniz means the active core of an animal machine, more akin to a software than to a hardware. When such a class of machines is complemented by the class of human-made machines, there results an enormous domain which deserves to be called a universe.

The adjective 'cybernetic' is justified by the definition of cybernetics as *the scientific study of the way in which information is moved about and controlled in machines, the brain, and the nervous system* (Longman, 1987), or – to complete the characterization – *the comparative study of the automatic control systems formed by the nervous system and brain and by mechanoelectrical communication systems and devices, as computing machines, thermostats, photoelectric sorters* (Webster III, 1971).

The core of this Leibniz's legacy can be best seen by comparison with Descartes who drew the sharp demarcation line between the automaton and the self-conscious mind. It was Descartes, and not Leibniz who throughout the three following centuries shaped the common view on mind, consciousness and matter. Only the second half of the 20th century witnesses the revival of the Leibnizian approach, due to Alan Turing, John von Neumann, Norbert Wiener (who suggested the term 'cybernetics' in 1947), and others. In Wiener's original statement, cybernetics is defined as 'the science of *control* and *communication* in the animal and the machine', to indicate (i) that the behaviour of the systems in question depends upon a flow of information, and (ii) that the laws governing control are universal, i.e. do not depend on the classical Cartesian dychotomy between mental and physical systems. This should not been interpreted as Leibniz's endorsing the materialistic monism which appears in some philosophical contexts of cybernetics (e.g., in the Marxist *Philosophisches Wörterbuch* (1969); philosophy may be combined with cybernetic approaches in different ways, and depending on the assumptions adopted, one obtains various philosophical conclusions (in Leibniz's case it was his infinitism which reconciled that approach even with theology).

The term 'cybernetics' became fashionable and made a brilliant career in the fifties and later. However, similar ideas appeared in other theoretical contexts, bearing different labels. Among the theories having similar objectives was general systems theory, information theory, and cognitive science. The last succeeded to become more and more influential. It can be seen as a generalization of what previously, since the sixties at least, was called cognitive psychology and was defined as a theory in which human beings and other psychological organisms are viewed as *information-processing* systems (Lycan 1990, p. 8). If the domain involved is not restricted to psychological organisms, and includes computers, robots, etc. as well, then the term 'cognitive psychology' is duly replaced by the more general one, viz. 'cognitive science' – to cover all possible cognitive systems, in particular those equipped with artificial intelligence.

In what follows, the basic term to be employed is 'information processing', and this allows us to see the present discussion as belonging to cognitive science. However, among the names of theories listed above, the only one from which a suitable adjective can be derived is 'cybernetics'. Fortunately, it is quite fitting because the controlled flow of information referred to in the definition of cybernetics amounts to information-processing, and that constitutes the subject-matter of cognitive science. Thus the universe being the domain of cognitive science can be referred to as the cybernetic universe, CU for short. It will be convenient to discuss the opposition of creative and mechanical reasonings with the help of this notion.

1.2.2. Here is the definition of CYBERNETIC UNIVERSE, given as a list of its constituents (in a fashion resembling a model-theoretical description).

The set of all individuals involved is the union of the three following classes: I, D and T, that is, Information pieces, Data, and some Things, respectively; the things to be taken into account are those which either process information pieces (as do, e.g., computers) or are processed by them (as are machine tools controlled by a software). Let x, y, z be variables to range over the whole union set; let i, j and d, e and t, u be variables to range over the component sets, namely I, D and T, respectively (if needed, in each category more variable symbols can be introduced with the help of numerical subscripts).

In T we distinguish the class of machines which is subdivided into natural machines (NM) in Leibniz's sense, i.e. those including organisms, and artificial i.e. human-made, machines (AM). Another distinguished subclass of T is that of Selves (S), i.e. self-conscious minds. The variables to range over these classes are lower-case letters hinting at the name of the category of question (n for NA, *a* for AM, and *s* for S), supplemented, if necessary, by numerical subscripts.

This list of sets is accompanied by a list of operations, called also functions. More exactly, these are classes of operations. For example, in the category PrD (Processing of Data) there is the subset of logical consequence operations among which there is, say, *ponendo ponens* as an individual operation (function); it assigns a sentence to a pair of sentences, both being classed with data. Before we discuss these classes of operations, let them be listed (in a perspicuous model-theoretical way), as follows.

CU = < I+D+M; NM, AM, S; Rec $[d,i]$,

Prl $[i,j]$, PrD $[d,e]$, PrT $[t_1, t_2, i]$, Cns $[s_1, x]$ >

In the first line above, the list of sets involved in CU is followed by the class of operations Rec, being fundamental for the whole construction. An operation of class Rec assigns a piece of data d a piece of information i (what is hinted at in square brackets following 'Rec'), so that the information piece is *recorded* by the datum. For instance, the same piece of information is recorded by each of the following pieces of data, belonging to Roman, decimal, and binary notation, respectively: [R] "I and II equals III", [D] "1+2 = 3", [B] "1+10 = 11". A closer examination of the relationship between data and pieces of information is found below, in Section 3.

The next three classes (the second line of the CU definition) involve various operations of processing, i.e. transforming something into something (the latter term is used, e.g., in the classification of syntactic operations into those guided by formation rules and those guided by transformation rules). The abbreviations hint at the transforming of: pieces of information into pieces of information, data into data, things into things (or, a state of a thing into its another state).

Things are processed either physically, as a piece of iron moulded by a hammer, or informationally, as a dog executing his master's commands. In the latter case, the content of the command is an information item, while the sounds produced to express the command are data, that is sequences of signs. Thus the dog undergoes transformation from a certain state (e.g., laying) into another state (e.g., jumping) what amounts to being in a way processed. Such a thing-processing is brought about by an element from category I, hence it is represented by the three-place function **PrT** $[t_1, t_2, i]$ to express that thing t_1 gets transformed into thing t_2 through a piece of information, referred to as i. Only this kind of thing-processing is found within CU, while physical tranformations (like that with a hammer) lay outside its boundaries (unless we take into account such a process as the influence exercised by the brain over the hand moving the hammer).

In the last mentioned class of operations, namely **Cns** $[s_1, x]$, the first argument belongs to S (self-conscious minds) and the other may be anything in our universe, as the mind is capable of being conscious of any object whatever (including its own acts; this kind of consciousness is called apperception, cf. Section 4.2 of this Chapter). To deal with this class is unvoidable in any discourse concerning reasonings, even if one may get rid of it when dealing with other parts of the CU realm, e.g., when constructing robots. A full artificial, i.e. mechanical, intelligence should involve elements of all the categories discussed above.

1.3. Information-processing through data-processing

1.3.1. Information pieces, or **IPs** for short, are entities thoroughly familiar to each of us; though, at the same time, they prove trickily elusive if one attempts to grasp their nature. In this category are found computer programs, meanings of utterances, industrial designs, musical structures embodied in score manuscripts, as well as human thoughts, either worded or tacitly entertained, and so on.[6] The first item mentioned in this listing, namely computer programs, belong to the category of SOFTWARE which deserves special attention both for technological and philosophical reasons.

Among crucial philosophical problems is that of interaction between the physical wòrld and the world of abstract entities (those discussed in passage 1.2.2 above). That interaction can be best studied in the case of programming languages of the lowest level,

[6]The class of **IPs** will prove more extensive if one adds genetic information and other factors controlling development and behaviour in the organic world. The problem of how to give a uniform and theoretically correct definition of such biological information pieces and such as those listed in the text above deserves a careful inquiry but, obviously, it cannot be included in the present book.

i.e., those in which expressions steering the behaviour of a physical system CD (ControlleD) are states of a physical system CG (ControllinG) such that they bring about some intended states of CD on the basis of physical causation. A paradigmatic example is found in the controlling of the movement of a vessel with its rudder. There is the abstract entity which we call *direction*, assigned to the vessel in question, and there is a set of abstract operations, viz. *changes of direction*, each of them being assigned to a physical operation which depends on a change of the position of the rudder with respect to the rest of the vessel. Thus a physical change of this kind is a word to express a command. Such a word exemplifies the lowest, that is, most direct level of software, devised for communication between physical systems. That the process deserves to be called communication is due to the meaning granted by abstract entities as assigned to physical events.

What all IPs have in common it is their relation to some physical objects termed as DATA; a piece of data will be called datum. The relation involved may be called expressing, representing, articulating, formulating, signifying, recording, etc., of IPs by data. Let the verb RECORDING be the standard one to denote this relation, while 'expressing' may be reserved for some psychological contexts.

Thus, an uttered (spoken or written) sentence is a physical event being a datum to record a proposition. A score, as a sheet of paper covered with notes (as material things made, e.g., from ink) is a datum to record a piece of music. A drawing is a datum to record the design, say, of a house. A configuration of polarized spots on a magnetic medium forms data to record, e.g., a program for computer.

To some extent, the elusiveness of the notion of IP can be remedied by a certain combination of the idea of recording relation with the concept of equivalence class (or abstraction class) as defined in logic and set theory. Let us recall that definition.

First, the EQUIVALENCE RELATION R on a set S is defined as one which is reflexive, i.e., xRx, symmetric, i.e., $xRy \rightarrow yRx$, and transitive, i.e., $xRy \wedge yRz \rightarrow xRz$, in S. For instance, the relation of being equidistant holds among the cities: Paris, Bordeaux, Lyons; that is, they have the same distance from each other. Whenever one has an equivalence relation on a set S, the set can be partitioned into a number of disjoint sets called EQUIVALENCE CLASSES such that all the members of any one equivalence class bear the relation R to each other but not to any members of S outside of that equivalence class. Thus the three mentioned cities (and, possibly, other ones) form an equivalence class included in a set of cities.

Now let us note the fact, which may seem curious but could hardly be denied, that to each equivalence class there corresponds exactly one abstract object; it will be called the object ASSOCIATED with the given equivalence class. If, say, A, B, C form the equivalence class of cities which are 200 miles distant from each other, then the abstract object associated with that class is *the distance of 200 miles*. In a sense, this 'extra' distance does not differ from the concrete distances between A and B, or B and C as each of them amounts to 200 miles, but it does differ in the sense that it is no concrete distance, as between, say, Warsaw and Cracow, but an abstract one in which concrete distances, so to speak, 'participate'.

If one wished to apply Ockham's rasor to this conclusion, by arguing that "entia non sunt multiplicanda praeter necessitatem", it should be responded that in this case we have to do with an invincible necessity of 'multiplying'. Provided the distance between Warsaw and Cracow amounts to 200 miles, we cannot help distinguishing between the concrete distance of these two cities and the distance of 200 miles 'in general'. The object *the distance between Warsaw and Cracow* is not identical with the object *200 miles* since there are predicates satisfied by the latter and not by the former. For example, it is true that "the distance of 200 miles was always the distance of 200 miles", while it is false that "the distance between Warsaw and Cracow was always the distance of 200 miles", as there was a time in which neither city existed.

On the other hand, we should not identify such an abstract distance with the class of equal distances coordinated to it, since the class of 200-miles-distances is not such a distance itself (the class has no geometrical dimensions). Hence, applying this conclusion to any equivalence class whatever, we are bound to state that with each of them there is associated an abstract object 'similar', in a way, to members of that class but not being one of them.

Before we demonstrate that IPs belong to that category of abstract objects, let other examples make this notion more conspicuous. Among concepts which often appear in everyday discourses are those of colour in general, and of particular colours as green, red, etc. Those things which have the same colour (this sameness being an equivalence relation) constitute equivalence classes, e.g., of green things, red things, etc. With each of such classes there is associated an abstract object, as greeness, redness, etc. They are necessary to sensibly answer such questions as "What *a* and *b* have in common?". In the case of green things, the answer reads 'greeness'.[7]

The same category of abstract objects involves numbers. Cardinal numbers are defined by reference to equivalence classes of equinumerous sets. E.g., number two is presented as the set of all pairs. This means that when resorting to such equivalence classes we win a reliable method to introduce numbers to the domain of arithmetic. However, this does not mean that the terms 'two' and 'the class of all pairs' denote the same entity. It is certainly true that *three is greater than two* but this does not imply that *the class of tripples is greater than the class of pairs*, hence the respective arguments of the predicate 'is greater than' do not prove interchangeable. Hence, again, a number as an abstract object appears as different both from the associated equivalence class and from its members.

There are well-known philosophical objections against acknowledging the existence of abstract objects, especially those different from classes (whose existence is granted by the axioms of set theory). However, for the purposes of the present discussion it is not necessary to settle this philosophical question. Acceptance of the category of such abstract objects as distances, colours, numbers, etc. can be merely pragmatical, that is motivated by a theoretical convenience. In the methodology of empirical sciences one is familiar with the notion of theoretical constructs. To refer in a theory to such constructs is one thing, and to have a philosophical belief that they do exist is another thing. The point of this chapter is that abstract entities of the kind discussed above are necessary for developing some useful theories, in particular IPs are necessary for a theory of reasoning.

---

[7]One can avoid the typical grammatical form for abstracts (marked by the ending '-ess'), and instead express the answer in a sentence like that: "They have in common that they are green". However, the phrase "that they are green" is only a different syntactic form to refer to an abstract object. It is a name which can be used as the grammatical subject in a sentence as, e.g., "That *a* and *b* are green is a desirable property".

1.3.2. The method of introducing abstract objects through equivalence classes, as exemplified above, should help to handle the notion of information, especially to examine relations between information-processing and data-processing. The idea to be developed is to the effect that pieces of information, abbreviated as IPs, are abstract entities, in a way assigned to respective equivalence classes of pieces of data.

Before we discuss the matter in a systematic way, some lexicographical report will be in order. Let us consider some samples of definitions which address the problem of relationship between 'data' and 'information'. For example, Greenstein (1978), a dictionary concerned with logic and computer science, offers the following explanation.[8]

"*Data* – A general term used for all facts, numbers, letters and symbols that are the basic elements of information capable of being processed by a computer." The term 'information' which occurs in this statement appears again in the following context. "*Data processing* – Any procedure for receiving information and producing a specific result."

The quoted statements exemplify two confusions concerning the term 'data'. First, it is not clear in them whether data belong to a language or extralinguistic reality. Letters and symbols, as listed by the dictionary, belong to the latter, while facts and numbers to the former (unless our author carelessly confuses numbers with numerals, and facts with sentences). Second, it is not clear whether information and data belong to the same domain or to different ones (even if coordinated with each other). According to the quoted dictionary, data are elements, i.e. parts, of information, hence they seem to be found in the same domain. Analogously, when the popular dictionary Longman (1987) explains the vernacular sense of *data*, identifies them both with facts and information, and when deals with a technical sense defines *data* as *information in the form that can be processed by and stored in a computer system.* Again, information and data are found in the same domain.

[8]To refute a possible objection that I make use of a suspect reference source, let it be noted that the quoted dictionary appeared with Van Nostrand, the publishing company renowned for its logic and computer science publications, and on the book flaps it is recommended as offering 'clearly stated definitions of technical terms'.

Even more professional Chandor (1985) lists numbers and alphabetic characters as two classes of data, hence fails to observe the difference between numbers and numerals, the distinction of domains to be crucial for a feasible definition of data-processing. Fortunately, in other entries Chandor (1985) provides us with useful hints to be developed in the present discussion. Here are his comments on the terms 'information' and 'data item'.

> INFORMATION Sometimes the following distinction is made between *information* and *data*: information results from the processing of data, i.e., information is derived from the assembly, analysis or summarizing of data into a meaningful form.

How it can happen that information so emerges from data, is a question which we are likely to successfully handle with the help of the following definition in the same dictionary.

> DATA ITEM A unit of data within an *application* system, one of the logical elements contained in a *record* and describing a particular attribute (e.g., name, address, age). May require a number of *characters, words, bytes*, or perhaps just a single *bit* to represent the entity concerned.

In the latter entry there are two expressions to lead our thought in a proper direction. One of them results from the idea to see the process of transforming data as a flow which needs to be segmented into items, hence the phrase *data item.* Owing to that, the paralell transformation process (going on in the counterdomain of a domain of data), viz. that of transforming entities represented by data, can be also divided into items; and thus a correspondence, crucial for the processes in question, can be established between them. The second fitting expression in Chandor (1985) is the predicate *to represent* which makes it possible to deal with the relation of correspondence between the two chains, that of data and that of entities represented by data.

The most basic is the correspondence between *numeral* sequences as data and *numbers* as entities *represented* by such sequences.[9]

---

[9]It deserves to be called basic (in a bit Pythagorean vein) since to any objects whatever one can assign numbers, and so numerals can be used to represent entities of any kind; even pieces of furniture, as well as their parts, can be numbered according to a convention, e.g. from left to right. The same procedure can be applied to abstract objects as well.

They belong to two different domains, each of them including objects capable of being transformed. E.g., owing to the successor operation, each number can be transformed into its successor (as 10 into 11), while in the counterdomain (i.e., the corresponding domain) of numerals, the sequence of numerals '10' gets transformed into '11'. The relation of representation comprises not only objects being transformed, and those resulting from transformations, but also transformations themselves, so that the transformation of '10' into '11' represents the transformation of 10 into 11.

These relations exemplify how a data item may represent an entity concerned, be it a number, a sequence of numbers, a function, etc. More generally, the item represented can be called an INFORMATION ITEM, construed as an abstract entity. Numbers are unquestionable citizens of the realm of abstract entities, but it may involve more kinds of them, as propositions, concepts, etc.

Data items are recorded in objects such as machines and organisms, and owing to these records information can be stored in objects and processed by them. To sum up this discussion, and to stress the abstract character of information, it is in order to quote the following definition found in literature concerned with foundations of computer science (e.g., in Turski 1985, p. 3). "Information is an abstract entity which can be stored in objects, transmitted between objects, processed in objects and applied to control objects, where by 'objects' we understand living organisms, technical devices and systems of such objects."

In the above definition, viz. in the saying that 'information can be processed', the concept of processing is applied to items of information, that is abstract objects, while in other contexts it is duly applied to data, that is physical objects. Obviously, these two processes are not independent; they are so related that data-processing is a *means* of information-processing. For instance, we transform the numerals '2' and '3' into '5' to learn from the result of that transformation that number 5 has been made from numbers 2 and 3.

This relationship between these two chains of transformations is the key fact to be taken into account in the further discussion. Let this be exemplified by the three following sentences, earlier discusssed in Subsection 2.2 of this Chapter. As marked by the prefixing letters, they are written in Roman, decimal and binary notation, respectively.

[R] "I and II equals III", [D] "1+2 = 3", [B] "1+10 = 11".

Let the term 'sentence' denote a physical object made from ink, or air waves, or electric impulses, etc. (while the terms 'proposition', or 'statement', or 'judgement' will never appear in this role). Thus sentences are found in the category of data. The sentences in question are so selected that they belong to a certain equivalence class; let it be called class E. Every sentence in E records the same IP. IPs recorded in sentences are called PROPOSITIONS, so sentences R, D, B record the same proposition; it will be referred to as P.

Obviously, P is not identical with any of the members of the equivalence class E. Neither with E itself. Were it identical with E, then it would be sensible and true to say, e.g., that P contains the empty class (as every class does) what, however, would be a kind of nonsense. Then, there are IPs which are abstract entities, each of them being associated with exactly one equivalence class of data.

To approach the problem of how information-processing is related to data-processing, let us first examine what means the former in the case of numbers. Numbers are IPs recorded by those data which we call numerals, e.g., written characters being names of numbers. Numbers can be processed with arithmetical operations. For instance, number 3995 can be processed with dividing it by 17 to result in number 235. It is usual to perform such information-processing by means of data-processing; the latter consists in writing down, e.g. with a pencil, strings of numerals and transforming them according to a certain algorithm. When doing that, one does not trace how the number itself is being transformed, he merely looks at the final result of data-processing to find that the associated information-processing results in number 235.

The above example is typical of what Leibniz called blind thinking (*caeca cogitatio*). A computing is not blind when one perceives numbers themselves, and how they change owing to applied operations; e.g., one is 'watching' how six objects, usually indefinite as to their properties, split into two halves, that is, are divided by 2, to result in tripples, which means that number three is the result of this processing.

The notion of blind thinking was meant by Leibniz as a more general term to cover both computing and reasoning, and possibly, other kinds of thinking. A part of this program has been instructively accomplished in our times as mechanical deduction carried out by computers.

1.3.3. In computing – to sum up – the domain of information-processing involves numbers and the domain of data-processing involves numerals. In reasoning, the former consists of propositions, the latter of sentences, and operations involved in data-processing are FORMAL rules of inference (in number-processing these are arithmetical operations recorded by such symbols as '+', etc.). The term 'formal' is to hint at their dealing merely with formal transformations, i.e. those concerned with the *form*, or structure, of strings of characters, that is data, and not with a content. About a reasoning which proceeds solely according to formal rules we say that it is FORMALIZED. Should such a reasoning be carried out by a machine, we call it MECHANIZED (in sense (ii) according to the distinction suggested above, in 1.1.2).

However, there are important differences between these two domains of information-processing which are crucial for our discussion (though not easily noticeable in Leibniz's time, when mechanical reasoning was but a visionary programme).

Data-processing in the sphere of computing, i.e. mechanical calculation on numerals as representing numbers is a usual and unavoidable procedure. Without it we could not handle even simple arithmetical operations (as in the example in the preceding subsection), hence it is a necessary and essential factor in mathematical activity. Had we not had these procedures, our capability of counting would not have exceeded that of paleolithic people.

On the other hand, data-processing in the sphere of reasoning, i.e. formalized inference, is a relatively new invention which has a clearly artificial character. Though it has proved necessary for metamathematical research, as well as useful and inspiring for philosophy of mind, it does not prove necessary for efficient reasoning. Albert Einstein could not do without arithmetical data-processing in his computations, but had no need to resort to rules of formalized deduction in his reasonings (even the mature methods of formalization are historically a bit later then early Einstein's works).

To illustrate this point, let us notice how often we reason quite instinctively according to the rule T (for 'transposition') which in a formalized logical system would read as follows:

from $(\alpha \wedge \beta) \rightarrow \gamma$ infer $\neg\gamma \rightarrow (\neg\alpha \vee \neg\beta)$

Here is the example of a reasoning carried out according to this schema.

(i) *Matter and its motion results in time and space.* Hence (ii) *If matter and its motion disappeared, there would no longer be any space or time.*[10]

Nobody needs to learn the formalized record of T in order to be capable of either inferring (ii) from (i) or acknowledging validity of this inference. Moreover, this inference corresponds to T only in a most rough outline; more rules, and more sophisticated than T itself, are followed in this inference. One assumes in it that the verb 'results' should be translated into 'if ... then', that the conjunction of names 'time and space' should be taken as abbreviation of a sentential conjunction, that 'no longer' is to express the denial, that the use of subjunctive mood should be subsumed under the logical form of implication, and so on. These additional rules are alien to formal and formalized logic, though quite familiar to the users of English.

This seems to be a convincing evidence that we need not formalized logical rules of data-processing in order to efficiently reason, either in science or in our everyday affairs. Instead, some other rules of data-processing are needed, and those are provided together with that know-how which is contained in our linguistic competence. To prove this point, a more elaborate discussion is necessary; it follows in the next Section which is to provide a broader framework to show the nature of mechanized reasoning.

1.4. Intelligence and model-based reasoning

1.4.1. Formalization, or mechanization in a broader sense (discussed above in passage 1.1.2 as sense (i)), constitutes the extreme at the scale of degrees of verbalization of inferences. In a formalized reasoning no premise and no consequence remains tacit, and each inference step has to be explicitly justified by a comment to refer to a relevant inference rule. Mechanization in the strict sense (see (ii) in passage 1.1.2) proves a suitable means to check whether the

[10]Sentence (ii) is one with which Einstein addressed in New York (1921) journalists who wanted him to explain in one sentence the basic tenet of his theory.

formalization in question is correct; a process of formalization when carried out by such devices as human eyes and a human brain may be not quite reliable, hence a severe mechanical test of correctness is highly desirable.

The class of formalized and mechanically checked reasonings constitutes the upper extreme of the scale both of verbalization and reliability. Below it there spans a wide spectre of reasonings which do not attain that high ideal. Some premises in them are assumed just tacitly, and some inference steps remain unnoticed. Some of them are nearer to the upper extreme, other ones are further, and so arises the question what should be found at the opposite extreme, i.e., the lower one.

(1) Are there entirely wordless reasonings, such that they do not resort to any piece of a text? (2) If there are, what, then, are the data to be processed? (3) What about the principle that every reasoning is a truth-preserving information-processing, where abstract pieces of information are represented by data as physical entities?

Suppose, the first question is answered in the affirmative. Then we shall need a handy term to call that lower extreme, that is, the class of reasonings in which information-processing is not supported by any text-processing. Let it be called the class of MODEL-BASED REASONINGS.[11] Another designation might be given after Popper (1982, p. 187) who observes *that we do think in words as well as by means of schemata.* The word 'schema' could give rise to the phrase 'schema-based reasoning', but 'model' seems to be a more convenient expression (for its use in the AI circles, for its having the verb variant 'to model', etc.). Now the first of the questions put above can be restated as follows: Are there any model-based reasonings?

The answer in the affirmative is supported by various kinds of evidence. Due to some natural laws governing organisms, people and animals are capable of forming, e.g., internal pictures of things. Such pictures and similar devices, some of them of more abstract character, can be called MODELS. This term enjoys sufficient generality and, like 'picture' involves the notion of similarity in its content.

---

[11]This term can be found, e.g., in documents concerning a workshop sponsored by the *American Association for Artificial Intelligence*, available in Internet (Gopher).

When observing behaviour of higher animals, we can explain it as processing pictures into other pictures according to some laws of logic. Any conditioned reflex, for instance, falls under the schema of *ponendo ponens*; one can also observe how an animal applies the procedure of eliminating members of disjunction. (Why not, if a computer can be 'taught' such procedures, while the nervous system of a roach is more involved than any computer circuits?)

Humans beings, in principle, enjoy the possibility of using language to express their reasonings. However, there is more information carried in the picture of a thing than in any verbal description of the same thing to be given at the same time, or even in longer time; there is, e.g., more shades of colour perceived by our eyes than words to describe them in a language. For instance, a fighter has too little time to produce a text which would describe all the moves of him and his rival to be taken into account in the reasonings which are to solve the problem 'how to win?'. Moreover, no vocabulary is sufficient to describe all the factors involved, as positions of fighting bodies, schemes plotted by one side and guessed by the other, etc.

The same reasons prevent a craftsman, or an engineer, from carrying out text-based reasonings. After all, why should he try to name all the states of the engine under repair, when he sees them with his eyes, and can test his unspoken estimations with movements of his fingers? His reasoning consists in transforming such visual and tactile data in his imagination; he is not bound to record them in his notebook and adopt predicate calculus for their inferential processing.

As for the reasoning of scientists (for whom Aristotle created his logic), let us note that typical scientific discoveries derive from thought experiments, not from manipulating verbal texts. It was a thought experiment of fancying a body that moves without being slowed down by friction which resulted in the following conclusion (the law of inertia): if no unbalanced force acts on a body, it will continue to move with undiminished speed in a straight line. This was Galileo's reasoning. Analogously, the anecdote of Newton's apple renders the fact that the discovery of gravitation needed an act of imagining behaviour of things in some supposed conditions, instead of studying and processing strings of written or spoken descriptions. And that is to say that Galileo and Newton, Descartes and Leibniz, etc., owed their achievements to masterful dealing with models. They were champions of model-based reasoning.

The second question posed above, viz. about the kind of data processed in model-based reasoning, should be settled as follows. While in a formalized reasoning, which is typically TEXT-BASED, the data processed are pieces of a text, in a MODEL-BASED REASONING the data processed are models, and those are due to records made in a code inside a processing system (e.g., a visual percept encoded in a brain). These records are also data to represent pieces of information, while in a text-based reasoning information pieces are represented by data formed as sentences. In both cases appear pieces of information of which either truth or falsity can be predicated (e.g., the truth of a percept involved in a model), hence the difference in their representing by verbal data in one case and model data in the other does not affect the nature of reasoning as a truth-preserving transformation; this is the answer to the third of the questions listed above.

1.4.2. Now we are to deal with what might be called a double 'mystery' of covert reasonings. To introduce this part of discussion, which should reveal a new feature of model-based reasonings, let us return to a methodological comment made at the beginning of this chapter, that about the process of merging logic with cognitive science. In understanding this process, mechanization of reasoning provides us with something like a contrast medium used in radioscopic examination: thus we can realize some radical differences between the mechanical reasoning as feasible for computers and the inventive human reasoning. In the example of mechanized inferences we best learn the core of reasoning which consists in preserving the truth (as in the mechanical procedure the truth-preserving rules are most explicit), and then we can search for the same core in those inference phenomena which in other respects contrast with mechanized reasoning, and should be treated rather by cognitive science than by traditionally minded logic. One of such differences, namely that mechanized reasoning is totally text-based while actual human reasonings happen to be model-based, was discussed above, and now it is in order to discuss another contrast.

The point to be vindicated is to the effect that there are reasonings which the reasoner is not aware of, i.e., those which do not occur at the level of conscious reflexion. We shall briefly speak of them that they are not *apperceived*, taking the term in its Leibnizian sense (in a way, being familiar to modern psychology too). Leibniz needed the word *perception* to denote all acts of living individuals (i.e., substances) as reacting to certain impulses, while fully conscious perceptions were by him distinguished with the specially coined term APPERCEPTION. He defines it as *the reflective knowledge of an inner state*, which is not given to all souls, nor at all times to the same soul.[12]

Now the point can be stated briefly that there are unapperceived reasonings in humans, and, still more briefly, that there are COVERT ones.[13] This statement is of great import for the study of intelligence. Provided it is right, the attempts to create artificial intelligence, which would be as close as possible to natural thinking, should not lead toward the text-based mechanical reasonings. Instead, artificial minds should be able to simulate model-based and covert reasonings as proving most efficient in those situations is which the subject of reasoning is not liable to be described in words, and, in addition, it has to be grasped in a fraction of the second (for instance, when a driver has to instantly find out the proper solution of a kinetic equation to avoid collision with another car). Thus a clear negative example, able to avert us from a wrong path, can be set owing to the theory and practice of mechanized reasoning.[14]

There is, actually, no mystery either in model-based or in covert reasonings, they are simply facts of every-day life. Nevertheless, there are philosophical schools such that one of them denies possibility of covert mental acts, and the other possibility of model-based

[12]See *Principes de la nature et la grâce fondés en raison*, Section 4, in Parkinson (ed.) (1973) p. 197. Cf. *Nouveaux Essais* [...], Preface, in op. cit. p. 153. As for the present use of the term, see Webster III.

[13]The term 'covert' alludes to the distinction found, e.g., in behaviourism, between overt and covert behaviour, the existence of the latter having been denied by radical behaviourists. However, in the present context 'covert' is to mean something different, namely the fact of being concealed even from the subject's own consciousness, not only from an external observer.

[14]Model-based reasoning is unvoidable in that kind of mental activities which is termed *knowing how* (in contradistinction to knowing that) by Ryle (1949), and more commonly is called 'know-how'. This problem was tackled by Herbert Breger in (Breger 1988) and other papers of him (cf. Leibniz Bibliography in *Studia Leibniziana*).

reasonings. From the latter point of view, that of behaviourists, there is a mystery in the conception that some reasonings might be non-verbalized since any thinking is by them construed as an inner silent speaking.

On the other hand, a covert reasoning is regarded as impossible by the Cartesian philosophy of mind in which the mind is identified with the subject of conscious acts. According to Descartes, there can be no covert reasonings, as reasoning is the affair of consciousness alone: I reason then and only then, if I know that I reason. There is no necessity for reasonings – meant Descartes – to be recorded in words (hence a model-based reasoning might be admitted), but it is necessary for them to be self-conscious.[15] This is why Descartes denied animals any capability of reasoning; he regarded them as mere automata so unable to make inferences as is unable, say, a clock.

The common-sensical point which does justice to our everyday experiences proves articulated in Leibniz's philosophy of mind. Leibniz applied his *lex continui* to this domain as well. Our ideas and, thereby, reasonings which involve ideas – claimed – can be more or less conscious, up to the degree which amounts to the lack of consciousness. This lack characterizes animal reasonings, but also many among those performed by men who, in principle, are capable of being aware of them but not always can, or must, exercise that capacity.[16] In this Leibnizian perspective reasoning is conceived as a kind of information-processing accomplished through data-processing, while the data are not necessarily linguistic; they may be some records in an organic machine which function as models of those pieces of reality which form the subject-matter of our reasonings.

Hence, there is nothing mysterious about the fact that a cat finds out a perfect solution of the kinetic problem of how to contract muscles in order to make a jump. An action of data-processing must proceed in the cat's body. In some conditions, when a human being is the agent in question, such a final conclusion may become

[15] An interesting comment on Descartes' radical rejection of language as a tool of reasoning is found in (Aarsleff 1993, p. 175 ff.).

[16] Leibniz's views in that matter are clearly expressed in his "Meditationes de Cognitione, Veritate et Ideis" in (Gerhardt 1880) vol. 4. Their modern interpretation in terms of cognitive science can be found in (Dipert 1994).

the subject of one's awareness, even without his being aware of the process which resulted in the conclusion. This is the typical case of illuminations having been experienced by those researchers who were lucky to make discoveries without being aware of inferences leading to them.

Both examples, that of a jumping cat and that of a suddenly illuminated scientist, illustrate what we have called a covert reasoning. However, there is that difference that for the cat the whole process is covert, while our scientist experiences the miracle of becoming aware of its conclusion without knowing premisses and intermediary steps. To support the point in question with a concrete example, the following story, once told by a logician reflecting upon his own mechanisms of reasoning, should be in order.

> Once, when I stayed in Salzburg, a town divided into halves by an oblong hill, on a July afternoon I went from my home to the central district. To reach it I had to cover – a several minutes' walk – a section of the road going in the tunnel under that hill. On that day, when entering the tunnel I left behind the clear sunny sky, but when I left it on the other side I was struck by the dark colour of the sky. I remember to have made two movements, not accompanied by any words in my mind nor even by any thoughts that could be singled out. I first looked at my watch and saw its hands indicating five minutes past five, and then without a moment of reflection I made about turn and went through the tunnel to that side of the hill from which I had just come. It was only on my way back that I realized why I had turned back and what was the connection of that with my having looked at my watch. After having noticed the darkened sky I must have thought (as I reconstructed that *ex post facto*) that it was already evening, which, however, was contradicted by the premiss that there was not a track of dusk when I was entering the tunnel, and dusk could not have come in a couple of minutes. Hence, in order to confirm that premiss (should my memory have failed me) I looked at my watch and made sure that it was in fact not a time for dusk. I then must have – I continued my reconstruction – acted under the influence of an alternative premiss (but I assure you that I had not even a trace of that premiss in my consciousness). It would have been the premiss "it is dark either because it is evening or because it is going to rain". My glance at the watch abolished the first constituent of the alternative, and hence

> the second one remained as the conclusion. That must have been followed by still several other steps in my reasoning (none of which, I stress that again, occurred in my consciousness), namely that it would not be good to get drenched, and since it is going to rain and I have no raincoat, getting drenched was quite probable, so it was worth while to avoid it by taking the trouble of returning home and taking an overcoat. *Here must have been such a reasoning* because it bore to my decision to go back the relation which the cause bears to the effect, but its occurrence was for me a *hypothesis* enabling me the comprehension of my own behaviour, and not an object of experience: I experienced no thoughts forming a reasoning, and even less so no thoughts dressed in words.

So far we used the case of mechanized reasoning as a contrast medium to show more clearly those features which are alien to mechanization. However, it is also possible to take advantage of the notion of computer as a positive hint to shed light at the phenomenon of covert reasoning. Namely, we distinguish between two processes involved in a mechanized reasoning: (i) that going on inside a computer, consisting of electrical impulses in a coded form (according to the machine code system of a particular computer), and (ii) that which consists in transcribing these internal data into data that can be read by human beings on an output device, e.g., on a screen where they can be examined visually.

The internal code is covert for a person operating the computer, while the data on the screen are overt; were the computer its own operator (as is each of us with respect to his or her brain), then the screen content would be comparable with what the operator is conscious of, while the record of the same reasoning in the machine code would remain hidden. In these terms, the story of a suddenly illuminated scientist can be rewritten as one in which just the conclusion appears on the screen whereas the rest remains in the machine code alone.

Even if the use of computers for making inferences is no breakthrough in our theorising about reasoning (Leibniz anticipated this long before computers), it is a significant heuristic means. It happens that some philosophers contest the theory of an internal code in organisms as something hardly conceivable and unscientific; the existence of such a code in machines should help them to reconsider the issue. Now we can define COVERT REASONING not only as a process which is just conjectured, and is described only negatively, i.e., by contrast with familiar overt inferences, but also as something which is known to humans as constructors of computing machines. In this way we win just an analogy, but a very instructive one. If someone denied such an analogy between machines and organisms, he would be burdened with defense of Cartesian position in the age in which biological science, unlike in the 17th century, can tell a lot about codes acting in organisms.

1.4.3. The main mechanical-intelligence problem can be stated in the terms of processing Encoded Potential Concepts (EPCs). That there do exist ENCODED POTENTIAL CONCEPTS in human bodies, is a philosophical hypothesis to motivate a research project. What EPCs are, should become more and more clear in the course of the discussion to follow, but a first rough explanation is due at the very beginning.[17]

An instance of EPCs on which attention of scholarly circles was focussed in recent decades is concerned with the notion of linguistic competence as introduced by Noam Chomsky; to start acquiring a language, a human being must have some innate potential notions of language, communication, predication, etc. Even if one does not endorse a materialistic point, it is advantageous to imagine those notions as encoded in our bodies, as a kind of data to be processed, presumably in the central nervous system. To use a classical example, we may refer to Plato's *Phedo* where Socrates ingeniously shows how a mathematical idea is found only potentially in a boy's mind, and then it develops up to the level of conscious and verbalized understanding (apperception, to use the previously introduced term). Though with Plato no biological interpretation is given, it does not contradict his system if one conjectures an idea to be recorded in a biological code – to wait, as if 'sleeping', for the moment in which it is 'read out', or 'decoded', i.e., processed and apperceived by the conscious mind.

Besides grammatical (referred to by Chomsky) and mathematical (referred to by Plato) encoded potential concepts, there are logico-ontological EPCs which prove indispensable at the start of

[17]Only few problems regarding the mechanization of concept-processing phenomena could be mentioned in this Section. Some other problems are discussed in Makowsky (1988) and still other ones in Marciszewski (1993).

any language acquisition. Among them are those of a *class*, of an *individual*, and of *equivalence relation.* True, they appear under such terminological guise at a sophisticated stage of language development; those who do not study logic or philosophy may never learn the terms, but the ideas themselves must be given to everybody who is able to learn language by what we call *ostensive definitions* (and there is no other way to start the learning of a native language). These three ideas are involved in any act of realizing that an individual object shown in the moment should represent a class to be named so and so, namely the class of those individuals which are identical – in a certain respect – with that being produced (i.e., an equivalence class). No communicative act involving ostension would be possible without functioning these logical EPCs, hence they must be innate in every human individual. Then, any artificial intelligence should have them inbuilt as well, to match natural intelligence in language acquisition.

There must be enormous, possibly infinite, resources of EPCs stored in each mind (this term – in the present context – is to denote the mind together with its physical setting). Only a tiny part of them may become deciphered, activated and developed in the course of an individual life. The great problem of strategy for an individual mental development is that of what should be so processed and by what methods. The same problem awaits the mechanical intelligence, provided that earlier a still greater challenge is met, namely that of equipping a mechanical mind with a store of EPCs comparable with that granted humans.

These preliminary explanations and the preliminary statement of the problem make it possible to raise the next question, which reads as follows: how the knowledge of mechanized reasoning may advance the so designed study of intelligence?

A mechanized reasoning with the utmost clearness reveals that any reasoning is a truth-preserving information-processing carried out by means of data-processing, the data being entities as physical in their nature as are electric impulses, or magnetized spots, while the property of truth-preservation is revealed in the explicit application of deductive rules. This paradigm allows us to draw up something like a questionnaire to examine other processes, to what extent they deserve to be counted as reasonings. Thus we can measure the distance to that ideal extreme case and, accordingly, make greater or less use of our experiences won in the process of mechanizing deductive reasoning, when trying to mechanize other intelligent activities.

To apply that methodological knowledge to those processes in which EPCs are involved, let us consider the following imaginary example. Once upon a time there must have been a while in which a primitive man discovered similarity among, say, the sun, the moon, and the trunks of trees, though the first is hot, the second cold, and trunks in many ways differ from celestial bodies. However, he discovered that all those things have in common that they are round. In a rather long evolution, that must have led both to the abstract mathematical concept of a circle and to the technological concept of a wheel. Thus there must have occurred an action of information-processing – from perceptions of the mentioned bodies to apperceptions of those sophisticated notions.

Does such a process belong to reasonings? That is to mean, does it possess the property of truth-preserving? What physical entities are involved in it as data which, after a suitable decoding and a new encoding, would be liable to mechanization? These are items in our questionnaire which we owe to experiences concerned with mechanized deduction. One must ask these questions if one plans a mechanical mind to match the human mind.

Our primitive ancestor, mentioned in the imaginary example, started from percepts which were true even if unspoken, as 'the sun is round'. Suppose, the percept in question was not verbalized. Then, its further processing, say an inference using it as a premise, is model-based instead of being text-based. Then, the data processed are not words but some codes recorded inside the ancestor's body. That data-processing which parallels and represents a piece of information-processing should result, say, in the concept of a circle, and this equals coming to a set of true conclusions, namely those geometrical statements which are involved in the concept of a circle (as analytic statements, they arise from such an analysis of the concept of a circle that it yields right predicates). Though such a process can hardly be counted as a deductive reasoning, nevertheless it is a reasoning, inasmuch as true premisses have been transformed into true conclusions.

As a model-based reasoning, it cannot be mechanized through arithmetizing a text, i.e., assigning numerals to its constituents, then coding numerals in binary notation, and then expressing those binary sequences of digits as physical phenomena in an electric circuit. The only alternative would consist in arithmetizing that biological code which corresponds to the model aiding the reasoning in question. There has to be something like that, according to the unquestionable paradigm of cognitive science that mental phenomena must have their counterparts in certain bodily elements.

Thus there appears a strategy for constructing mechanical intelligence which would really match the natural intelligence of human beings. One should (i) discover mechanisms of model-based reasoning in order to imitate them with artificial devices, (ii) furnish such devices with a set of Encoded Potential Concepts similar to that enjoyed by humans, and (iii) master the process of transforming unconscious, only potential, ideas into fully apperceived concepts.

In this enterprise, a specially difficult problem would consist in dealing with vagueness and confusedness which are essential traits of EPCs. These seem to be represented by non-discrete, i.e., continuous quantities, while arithmetization is a discrete procedure. In other words, rather analog machines than digital machines seem likely to deal with such simulations. Obviously, we have learned to solve the problems of analog-digital conversion, but first the analog reality should be adequately recognized, and this is the greatest challenge as far as internal states of human bodies are concerned. So far, we do not know at which degree of complexity the codes in question are to be looked for, be it the molecular level, or be it, say, the level of quantum phenomena (cf. Penrose 1988, p. 514; Penrose 1989, p. 400, Davies 1993, p. 251), or else something between.

In the face of such enormous difficulties, we should modestly start from a careful phenomenological description of model-based reasonings and of the processes of transforming potential concepts into apperceived concepts, and so to prepare the base for a future exploration of physical counterparts of these mental processes. This should be a suitable job for mind-philosophical logic.

The success of mechanization of deductive reasoning which resulted from the accumulation of logical discoveries and technological inventions was unthinkable yet a century ago. Is there a chance to attain a comparable success in mechanical transforming encoded potential concepts either into apperceived concepts, or in unconscious, yet efficient, actions? In such a research, both a technological success and a technological failure would mean a cognitive success. The failure would show the human mind as being too great to be matched by any technological product of its own.

CHAPTER TWO

THE FORMALIZATION OF ARGUMENTS IN THE MIDDLE AGES

2.1. The contention of the present chapter

2.1.1. Why the story of mechanized arguments should start from the Middle Ages? One reason for doing so is that a good many of the storytellers begin their narration with Raymond Lull (1232/3?–1316), or Raymundus Lullus (the latinized form of the name), a Catalan schoolman active in the fertile border area of Latin, Arab and Jewish culture. The second is that those tellers are wrong, hence their mistake should be rectified. The third reason, the most serious one, is that mathematical machines in general, and mechanized reasoning in particular, have become possible owing to three historical achievements, namely the invention of electronic devices, the algebraic approach to logic, going back to the 19th century and anticipated two centuries earlier, and the creation of formalized languages, that goes back to a camp of medieval logicians (which, however, did not include the famed Lull in its ranks).

Why the last named factor is so important? The key notions in the story, as claimed in the preceding chapter, are those of information-processing and data-processing. The former can be entrusted to machines provided that information items (abstract objects) are represented by data (physical objects), hence the results of data processing due to a machine can be read off by a human as results of information-processing. In a process of reasoning, the formalization consists in recording propositions (information items) as data, and in putting forth inference rules as operating on physical objects. Thus formalization procedures contribute to the theory and practice of information-processing.

There are authors who ignore that vital connection between mechanization and formalization, and those are inclined to praise Lull as a forerunner of mechanization, though no traces of formalization appear in his doctrine and method called by him (without excessive modesty) the great art – *ars magna*. The opinion that Lull's *ars magna* made him the forerunner of the present methods of mechanizing arguments is being spread even by some recognized experts in field. Not only we find it in studies concerning history of computer science (e.g. Gardner 1958) and in serious works on the history of logic (e.g. Bocheński 1956, §38); also by, e.g., Bonner (1985, pp. xii, 312, 575), an expert on Lull's works and their translator into English, Lull is merited for the discovery of the nucleus of the future computer science.

The present study is intended to provide arguments which call that opinion in question. That unambiguously negative result becomes possible owing to the fact that a historian of logic working in the second half of the 20th century has a clearer idea of the mechanization of arguments than his predecessors might have had. He has in his field of vision the formalist experiments of Hilbert's School, the problems of decidability, the concepts of automaton, algorithm and Turing machine, and also a practical experience with the use of machines in automatic theorem proving. This is why he can better understand questions to be posed to Lull, namely those concerning the way and the scope of the mechanization of logic.

That result, which denies the alleged Lull's contribution to mechanizing reasoning, due to a better understanding of the concept of mechanization, increases, in turn, the degree of preciseness of that concept. For, by such criticism we contribute to making the concept in question still more precise. In the study carried out in the present book, we are in a position to compare a spurious positive case (Lull) with a genuine positive case (Leibniz), and also to hint at historical sources of the latter. It turns out that those sources are to be found in the formalist and finitistically-combinatorial tendencies which marked the principal trend in the development of syllogistic, from Aristotle to the *via moderna* in the late Middle Ages, marked by the nominalist orientation. This perception of continuity of development, instead of a search for a flash of genius on the part of the forerunners (which reminds one the search for the miraculous in history, criticized already by Descartes), is an important lesson in history.

Originally the present study was intended to present and, possibly, to press the arguments for Lull's being ahead of his times in the process leading to the mechanization of arguments. Yet a more careful examination of Lull's texts resulted in rejecting that prevailing opinion. Here are some results of that scrutiny.

What is taken to be the algorithmic component of Lull's Art, namely the mechanical formation of propositions by combining terms, ends in the formation of a sentence, while the decision whether that sentence is true or not is entirely left to intuition; and, to make matters worse, that concept of intuition is tinged with mysticism in the Neo-Platonic and Augustinian style. Hence Lull's candidacy to the status of the principal predecessor of Leibniz and the initiator of the mechanization of arguments proves untenable. Such is the result of the inquiry to be reported in this chapter. It is true that Leibniz had predecessors in the Middle Ages, but these are rather to be sought in the trend that was in opposition of the Neo-Platonic orientation represented by Lull, namely in the nominalistically-oriented *logica modernorum* in the late Middle Ages, also called terminist logic because of its semiotic inquiries into the so-called *proprietates terminorum.*

2.1.2. The stubbornly supported legend about Lull as the forerunner of the mechanization of arguments is the more astonishing in view of the fact that there are available highly competent studies which deny it. The most authoritative in that respect is the work of Risse (1964) which discusses, among other things, the Lullist tradition in the 16th and the 17th century, and hence in the perspective of the times in which it can be seen against the background of the more advanced plans of the mechanization of logic, in particular that of Leibniz.

Although Risse's excellent work appeared three decades ago, the myth of Lull has not lost its vitality, and still requires a polemical discussion. Hence a longer quotation from (Risse 1964, an *ad hoc* translation from German) will be in order to make account of the genuine approach of Lull.

> Unlike all other logicians, Lull and his followers construe concepts not as formal designations of things but as names given them by God which should reveal their essential properties. Hence, in order to acquire knowledge one need not analyse things in the light of experience but has to obtain insights in the underlying names of things, so-called principles, and to mutually relate these names. It is owing to the symbolic

> names and their signs that Lull's *ars magna* acquires its focal position in knowledge, a position which is in fact metaphysical and not logical.
>
> The Lull school is to be understood solely through Lull, not by reference to Leibniz. True, Leibniz owes many particular ideas to it but he organizes them in his own way. For with Lullists logic was neither primarily nor essentially connected with mathematics [...] (Leibniz's) *calculus ratiocinator* rooted in Vieta's *algebra speciosa*, is a product of the 17th century, and *ars magna* played no noticeable role in it. (p. 532).
>
> Lull's *ars magna* is mystically oriented towards the super-rational, not logically oriented towards the rational. For, unlike Leibniz in the 17th century, in the 13th century Lull does not primarily care about the mathematical problem of how to find the exact number of possible sign combinations, nor even about a mathematical calculus as a substitute for natural thinking, hence also not about how to logically solve the problem posed. Instead, Lull, having faith in the magical cognitive force of words, wishes perfectly to grasp the manifold properties of Being through combining concepts-principles which denote the essence of things. Moreover, Jewish mystics in their cabala, referred to as Lull's source, had rendered the essential divine attributes by concentric discs and by turning those discs they wanted to arrive at the many divine names. [p.533f; this sentence is completed with a footnote which mentions Abraham Abulafia (1240–ca. 1291) as Lull's cabalistic source and refers to (Schoolem 1955, p. 133 ff).]

The statements quoted describe that historical state of things which will be documented below with Lull's own texts. Risse's account is of particular importance; it shows that Lull was not even the author of the idea so often ascribed to him, that of the mechanical combination of concepts – the idea which, according to some simple-minded commentators, was to be a prototype of a logical machine. It turns out that the idea was first born in a context which had nothing in common with any logical theory. A more penetrating study of Lullistic texts allows one to note that the joining in them of a logical terminology to that apparatus used for the combination of names does not contribute anything to logic itself, and the appeal to intellectual intuition, so typical of Lull, separates those elements of logic from the formalist orientation, so fertile in the case of many medieval authors, namely the orientation towards logical form and algorithms.

2.1.3. The said orientation marks one of the remarkable stages in the history of formalization, and hence mechanization, of arguments. It deserves being recorded even though it could hardly be regarded as a turning point in the history of our problem. The real turning point was so thoroughly forgotten that the degree of its oblivion proves as great as the degree in which the legend of Lull's key role is spread. That turning point is mentioned by Risse in the second paragraph of the passage quoted above. It consisted in the emergence of algebraic symbolism as using variables. Special services in the field are due to François Viète (1540–1603), with whom we associate the concept of *algebra speciosa*, also called *logistica speciosa* and *analytica speciosa* (the last term was used by Leibniz in his introduction to *Dissertatio de arte combinatoria*, item 6). In each of those terms the noun denotes calculus, and the recurrent adjective *speciosa* derives from *species* in the sense of a set, and indicates the occurrence of variable symbols which refer not to numbers but to sets of numbers (*speciosa* also means 'beautiful', from *species* in the sense of 'face', and it perhaps suggested that the mathematicians in that epoch appreciated the beauties of the young branch of mathematics, but the pride of place goes to the technical, and not the lyrical, meaning).

The opinion that Renaissance mathematicians invented variables must be taken with a grain of salt. It is not groundless to ascribe that idea to Aristotle as the author of syllogistic schemata. Others, for instance A. N. Whitehead, ascribe the principal merit to Archimedes. When it comes to such a sophisticated concept one may assume in advance that it was developing for centuries stage by stage, and whichever stage is taken to be the turning point in that process, the decision will always be arbitrary. In any case, the introduction of variables in the 16th century and their intensive use by Descartes (whose merits are emphasized by Leibniz 1646) coincided with many other important inventions in mathematical notation as done by Cardano, Tartaglia, Stevin, *et al.*, and thus that notational device not only greatly improved the technique of computation but also inspired the idea of a universal formal language.

It was mainly Leibniz who was inspired by it. Combined with his acquaintance with the achievements of scholastic logic, which had already arrived at an advanced concept of logical form (see below in this chapter), that idea gave birth to another one, namely that of formalized logical calculus. This is not to say, however, that the idea of formalization and mechanization of logic was given by Leibniz a strong push in the sense of historical results. For his ideas and results remained unknown to his contemporaries and the posterity. Until the early 20th century, Leibniz had to play the role of the great witness of his times, whose testimony is instructive to our generation, but not the role of a great inspirer, which he could have played had he not buried his drafts in the archives of the Hannoverian library, incessantly postponing their publication in the expectation that he would obtain more mature results. Thus, independently of Leibniz, many authors in the 18th century strove for an algebraic formalization of logic (see Chapter 4), but only those who lived in the 19th century succeeded fully in solving that problem (see Chapter 5).

In that historical perspective the achievements of the medieval formalists do not appear to have been a breakthrough, but they did mark some steps in logic's march to its mechanization. In any case, they were real attainments, unlike Lull's doctrines to which legend has ascribed the merit of an early anticipation. That is why they are recalled at the very start of our story as a constrast-making background; this should help in the demythologization of Lull, and so contribute to the drawing of a more faithful picture of the historical process under study.

2.2. Heuristic algorithms in the Middle Ages

2.2.1. Paradoxically enough the problem of a certain formalization of arguments was for centuries discussed in terms of the logic of discovery, and hence in terms of processes which we now treat as typically creative and not subject to mechanical procedures. That was due to a combinatory interpretation of the process of discovery, which involved finitism and formalism. Finitism because effectiveness requires that the number of combinations be finite. Formalism because combinations must be carried out on some discrete object that can be unambiguously identified; these conditions are in a model way satisfied by material symbols owing to their visible shapes (or forms, hence the term 'formalism') and discrete arrangement. That finitistic formalism started in the late Middle Ages and culminated in the 17th century.

Before one analyses in greater detail those plans of the mechanization of arguments, it is in order to hint at the position of the theory of reasoning in the structure of traditional logic.[1] This structure is accepted by Lull and other schoolmen as well as the logic of Port Royal (though the latter was deliberately anti-scholastic in accordance with its Cartesian vein). The ordering schema was tripartite, and it started from singling out three hierarchically arranged operations of the mind.

Fundamental among those three operations was that of grasping things by *concepts.* In a sense it could be called the simplest one, and so it was called by scholastic logicians, with Lullists among them, when they used the term 'simple apprehension' (*simplex apprehensio*). It is the simplest one in the sense that in that part of the act of grasping things which occurs at the level of consciousness, and so is accessible to introspection, we do not perceive any components that could be clearly isolated.

For instance, some simple apprehension leads one to the concept of natural number, whose content includes, *inter alia*, the fact that every natural number has its successor. Obviously, no one knows what and how complex program is recorded in the "internal code" of the machine which the human nervous system is, namely the program which "on the screen of consciousness" produces the idea of an infinite sequence of numbers. We cannot, however, dismiss the supposition that such a program as well as the process controlled by it must be something vertiginously complex; hence, calling its results a simple apprehension may derive from one's being unaware of that hidden complexity.

This leads us to the first critical comment addressed to Lull and the Lullists, including the most eminent among them, namely Giordano Bruno. When treating concepts as something ready-made and

[1]The term 'mechanization' will be used throughout this chapter and later both in the narrower sense, to denote reasoning carried out by a physical machine, and in the broader (even a bit metaphorical) sense to stand for any mechanical procedure in reasoning, i.e. one that for which there is an algorithm (conceived as an abstract machine). In each case the context should prevent misunderstandings (cf. passage 1.1.2 in the preceding Chapter).

simple as to its origin, and as something to be taken as the point of departure in reasoning, they disregarded that vertiginous complexity of the process whereby concepts develop. That complexity cannot be simulated even by present-day computers nor deciphered by the sophisticated methods of present-day neurophysiology.

2.2.2. The second operation of the mind, second in the sense that it assumes the existence of concepts and is more complex than conceptualization, is the formation of *judgements*. According to Lullistic logic, and scholatic logic in general, it consists in the combination of concepts into a judgement (*judicium*). This had suggested to Lullus the idea of a combinatorial procedure of generating judgements. From the point of the present-day interpretation of language it was an Utopian plan and hence difficult to understand (if the comprehension of a given author is based on the assumption of his rationality). This is so because there is no limit of the complexity of sentences, even those which we term atomic, because in the construction of sentences we admit predicates with an arbitrary number of arguments; and the imposition of any constraints upon the number of arguments would either be conventional or appeal to intuitions that are far from being mechanizable.

In order to understand some motivation on which the programme of Lullism was based we should take into account the conception of judgement underlying syllogistic. A judgement was treated as an invariably tripartite structure consisting of the subject, the copula, and the predicate, the copula expressing either affirmation ('is') or negation ('is not'). This is how a judgement (protasis) was understood by Aristotle (see Prior Analytics 24a; cf. Bocheński 1956, items 10.07 and 12.04). For a properly limited dictionary of terms (i.e., expressions which can function as either the subject or the predicate) this yields a realistic algorithm to produce the list of all possible judgements. From such a list one would then have to choose those judgements which are true and as such are suitable as premisses of those syllogisms that are to yield truth. Today we find it difficult to imagine what algorithm which is not a proof (because the proof, e.g. a syllogism, comes later as the operation that is next to the formation of judgements) could be constructed for that purpose: empirical truths are beyond the reach of any algorithms whatever. The point is, however, that for the representatives of the Platonic-Aristotelian views (which came to be opposed by modern empiricism) truth in the strict sense of the term was identical with necessary truth or (approximately) analytic truth, and the set of such truths was supposed to be decidable. This might have resulted in a path that would lead to the algorithms of discovering the truth, that is, the glorified logic of discovery, *ars inventionis*, which preoccupied many outstanding minds in the Middle Ages and the Renaissance until the 18th century.

2.2.3. The third operation of the mind, the most complex one in the sense that it assumes the two preceding ones and brings the most complex product, consists in the *proof*, construed in the Aristotelian logic as a *syllogism*. This identification of proofs with syllogisms in that tradition is essential for the present discussion because syllogistic combinatorics, which is the foundation of mechanization can then be identified with the theory of proof taken as a whole, which would allow one to conclude that all proofs can be mechanized. Here is an account of the relation that a proof bears to a syllogism according to Aristotle.

> By a proof I mean a syllogism which creates scientific knowledge, that is such a syllogism owing to which, if we only have it, we possess that knowledge. Hence if knowledge is such as we have stated, then the premisses of demonstrative knowledge must be true, primitive, direct, better known (than the conclusion) and must be its cause [...] A syllogism can develop also without those proper premisses, but a proof cannot, for it will not produce scientific knowledge. *Posterior Analytics*, 7lb.

According to the above text the difference between a proof and a syllogism is epistemological and not formal logical in character: a proof is a syllogism whose premisses meet the epistemological conditions (specified in *Analytics*) of scientific knowledge (*episteme*) as distinguished from common belief (*doxa*). Hence in science there are no proofs other than syllogisms, even though not every syllogism is a proof. Thus *syllogistic* exhausts the whole formal logical part of the theory of scientific proofs to which, in Aristotle's intention, his *Prior Analytics* (concerned with formal logic) and

Posterior Analytics (concerned with methodology of science) were dedicated.

Someone who knows the history of Greek and later mathematics might object the point that logicians of that time identified a proof with a syllogism. They must have known the procedure employed in proofs by Euclid and other mathematicians which hardly resembled syllogistic forms. If so, why did they claim that every proof should be a syllogistic inference? And yet they did claim, and this lasted from Aristotle to Christian Wolff (1679–1754). The latter in his very popular textbook of logic (Wolff 1712 whose title in English would run *Reasonable Thoughts Concerning the Powers of Human Understanding and Their Correct Use in the Cognition of Truth*) firmly opposed the Cartesian view that syllogism is no proper tool in mathematical reasonings, and in his polemic with that view he tried to reconstruct, by way of example, Euclid's proofs with means which (in his opinion) were purely syllogistic.[2]

2.2.4. Nowadays it is well-known fact that syllogistic can be interpreted in the monadic predicate calculus which is a decidable theory. This fact may have been intuitively sensed in practicing syllogistic inferences, and together with identifying the whole of logic with syllogistic that might have led to the belief in the possibility of mechanizing all reasonings. This belief seems to have been favoured by other factors in the cultural context in which logic existed for two millennia. That context involved two philosophical tendencies, namely finitism and formalism.

In the Greek philosophy of mathematics *finitism* established its place for good owing to the paradoxes of Zeno of Elea (490–430 B.C.), which showed how formidably perplex are the problems resulting from the concept of (actual) infinity. That was reinforced by the authority of Aristotle who in his *Physics* and *Metaphysics* (e.g., 1048b, 1066b) advanced arguments, to be later repeated for centuries, against the existence of actually infinite domains. That

[2]In his youth Wolff was a firm adherent of Descartes whom he followed in fighting against syllogistic as a method of mathematical reasoning. Later he was convinced in a correspondence with Leibniz that the formalism of syllogistic should not be disregarded. Unfortunately, he did not follow Leibniz's attempts to develop that formalism toward a mathematical form, and believed in a proof-theoretical omnipotence of traditional syllogistic.

opinion was also represented in antiquity by other schools but Aristotle's voice sounded more loudly, especially when supported by Christian thought: the major part of its representatives reserved infinity for the Creator alone while denying it to the creature. The finitist camp included such influential authors in Christian antiquity as Origen (185–254), with whom centuries later Cantor himself would engage in vehement disputes, Proclus of Constantinople (410–485), an influential commentator of Euclid, and – in the period of the flourishing of medieval philosophy – Thomas Aquinas (1225–1274), the greatest Christian Aristotelian. That camp did not include Augustine of Hippo (354–430), but his standpoint, voiced only in connection with other problems and hence likely to be overlooked, was fully understood probably only by Cantor (who sought in him as ally in polemics with contemporary theologians).

Thus the intellectual climate in which the Lullists were active favoured finitism. The view which implied the finiteness of both the domain of individuals and the set of concepts provided people with reasons to postulate the decidability of the system of human knowledge. True statements would be deducible from a finite set of first principles (as Aristotle claimed), and false ones would be refutable by demonstrating that they contradict those principles. It can thus be seen that the propagation of the idea of the decidability of all problems required of Lullus neither any exceptional originality nor the overcoming of resistance on the part of his contemporaries. This fact to a large extent explains the popularity of Lullism and its expansion during several centuries.

Likewise, a kind of *formalism* was at that time nothing extraordinary. Its lively presence can be seen in both the intensification of the nominalistic trends from the 12th century to the late Middle Ages and the 17th century, and also in other elements of medieval culture. It was not without importance that human minds were at that time imbued with biblical ideas, which included the faith in the perfection of the concepts to be found in the Scripture and the adequacy of the words which rendered those concepts. For instance, since Adam in Paradise gave names to all animals, that is, came in the possession of their concepts, nothing more was left in that respect to be done. No new intuition was necessary, it just sufficed to make use of the words assigned to the concepts. Likewise astronomy was a complete and closed system, which was moreover perfectly synthetized with the theological one (which we can finely seen, for instance, in Dante's *Divine Comedy*). The ideas of the potential infinity of human cognition, of the limits of verbalization, of the approximative nature of scientific theories, and the like, have become familiar to the modern mind only recently. As long as it was believed that concepts and judgements had adequate mappings in language, on the one-to-one basis, there were reasons to believe that thoughts could be fully replaced by words, and these, being material objects, could be processed mechanically.

This picture of medieval mentality applies to only some of its trends (it would be naive to treat it as a monolith). For instance, Augustinism opposed Aristotelianism both by its infinitism and its doctrine of illumination, which stressed the intuitive, non-mechanizable elements of cognition; that tendency even more manifested itself in the gnostic movements. But as for the problems with which we are concerned here the essential point is that both the finitistic and the formalistic trend were firmly rooted in the medieval thought.

2.2.5. After having outlined in 2.2.3 the notion of proof in traditional logic, and in 2.2.4 the intellectual background that favoured the idea of the mechanization of arguments, it is in order to trace the history of that idea in order to weigh with justice contributions of medieval logic and innovations due to Leibniz and others. As in so many logical issues, the origins go back to Aristotle. It was he who raised the problem of finding fitting premisses from which an intended conclusion should derive. (cf. Bocheński 1956, Sec. 14.29). A large passage in *Prior Analytics* (Book I, 43a ff) is concerned with that question. The passage opens with the following announcement.

> And now one should say how we arrive at possessing syllogisms which are always suitable for a given problem and how we obtain the principles proper to it (i.e., a given problem). For we should perhaps not only reflect on the structure of syllogisms, but also have the ability to form them.

That passage gave rise to some problems handled in late antiquity (Joannes Philoponus, 6th cent.), and intensively analysed in the Middle Ages under the name of *inventio medii*, i.e., the problem of finding the middle term. It was typical of *logica inventionis*, that is, the logic of discovery, postulated also by Lull, who wrote, among other things, the treatise entitled *Ars inventiva veritatis*. Averroes (1126–1198), one of the greatest Arab Aristotelians, was the first medieval author known to have coped with the issue. On the other hand, Albert the Great (1193–1280) was the first who, following Averroes, assimilated that problem to Christian scholasticism. Albert's another merit consisted in that he took up, after the Arabs, combinatorial investigations concerning the number of all possible syllogistic structures in order to find all possible correct syllogisms. A similar combinatorial approach can be found in the Jewish logician named Albalag, who was a contemporary of Lullus and like the latter was active in northern Spain. As can be seen, when it comes to the combinatorial approach, Lull was by far not the first among the schoolmen, and if he did not take that problem for instance from Albert the Great, then he must have most probably owed it, as Albert did, to the Arabs.

Among the authors who took up those problems after Albert the Great special mention is usually given to George of Brussels and Thomas Bricot.[3] They belonged to the nominalistic trend in the 15th century (*logica modernorum*), which had originated with the great Oxford masters, William of Ockham (1300–1350) and Richard Suiseth, called Calculator, the author of *Liber Calculationum* (1328), whose teachings transferred to the Continent by Paul of Venice (d. 1429) took strong roots at the universities in northern Italy and central and eastern Europe, including Prague, Cracow, and Leipzig. The last-name place is of particular importance for our story because it takes us to the reading list of the young Leibniz who, before having written his *Dissertatio de arte combinatoria*, whose subtitle refers to *logica inventionis*, had literally within the reach of his hand numerous dissertations on similar subjects treated in the spirit of nominalistic logic, which circulated in that milieu. It was Leipzig, too, which saw in the early 16th century the appearance of Gregor Bredekopf's *Tractatus de inventione medii*.

2.2.6. Thomas Bricot, mentioned above, in his commentary to the lectures to George of Brussels presented the teachings on how

[3]See (Bocheński 1956, Sec. 32.36) and (Swieżawski 1974, vol. 2, IV.2).

to find the middle term; it is worth noting that both, the lectures and the commentary, were very exceptionally widely read. Here are examples of rules for that operation.[4]

> In order to prove a general affirmative sentence (SaP) one has to take such a middle term M which follows from S and from which P follows.
>
> In order to prove a particular affirmative sentence (SiP) in the Darapti, Disamis, and Datisi moods one has to take such a middle term from which S and P follow.
>
> In order to prove a general negative sentence (SeP) in the Celarent and Cesare moods one has to take such a middle term which is mutually exclusive with P and follows from S; if the proof of SeP is to be in the Camestres mood, the middle term should be mutually exclusive with S and follow from P.
>
> In order to prove a particular negative sentence (SoP) in the third figure one has to take such a middle term from which S follows and which is mutually exclusive with P.

The full set of such rules in the form of a graphical schema, nicknamed *pons asinorum*, was given by Petrus Tartaretus (a well-known Scotist and rector of Paris University in 1490) in his commentary to Porphyry's *Isagoge* and to Aristotle's logical writings (Tartaretus 1517); the diagram itself is supposed to have been constructed ca. 1480. The term 'bridge of asses', to this day preserved, oscillates between two meanings. Tartaretus himself, in his desire to prevent his students from experiencing anxiety in view of the complexity of the diagram, explained that such anxiety would be as groundless as that which is felt by asses which are to enter a bridge, because that diagram is to ensure a safe passage, and not to render it difficult. In the other meaning, until today to be found in dictionaries (e.g., Kondakov 1978), one takes into consideration the feature of facility which marks algorithms, that is, mechanical procedures that do not require inventiveness on the part of the user, and as such are manageable even by proverbial asses.[5]

---

[4]Cf. Swieżawski 1974, pp. 174 and 186. The formulations quoted are paraphrases of Bricot's text *Cursus optimarum quaestionum* into German to be found in Bocheński (1956, Sec. 32.38) who availed himself of quotations found in (Prantl 1870, vol. 4, note 129). Bricot's rules do not constitute a genuine algorithm but are some steps in this direction.

[5]This bridge of asses is not to be confused with the so-called Buridan's ass. Buridan, one of the most prominent logicians in the first half of the 14th century,

The picture that emerges from the foregoing overview does not confirm the opinion about Lull's role in shaping the idea of the mechanization of arguments. The key role was played by late medieval nominalists with their definitions of logic as pertaining to terms, and hence physical objects on which mechanical operations can be performed (note in this connection that the leading nominalists, such as Ockham, Suiseth, and Buridan, were also forerunners of modern mechanics). It was nothing else than terministic logic, that *logica modernorum* which formed the main link between Aristotle (whose texts admit a "terministic" interpretation but do not forejudge it) and Leibniz, who, after having found himself, together with his Leipzig masters and Thomas Hobbes in the sphere of nominalistic inspiration, transformed it in his vision of logic embedded in a mechanism.

2.2.7. How should we assess the movement in favour of *logica inventionis* from the point of view of the present-day logic? Such an attempt should take into account those vehement polemics with Aristotelian and scholastic logic which took place in the Renaissance and the 17th century. When comparing their arguments with our present knowledge, we find premisses for a well-balance evaluation.

What strikes us today as the value represented by the formalist wing of that logic, namely the approach to possible mechanization, was severely criticized, at first by numerous humanists, who postulated that logic should "come closer to life", the postulate marked by their practical and psychologistic attitude, typical of rhetoric. At the same time there was criticism in the vein of Francis Bacon, which demanded that logic should discover hard natural facts and not mere terms (such as the middle term to be found with medieval recipes).

In the 17th century, in turn, the formalistic trend had its main opponents in Descartes and his followers, who ascribed to logic the role of the healer of minds. According to that programme logic was to protect human minds against deviations, including the scholastic and formalistic ones, and to make them capable of

being a leading nominalist, also appreciated logical algorithms, but he is said to have used the picture of ass for other purposes, namely in a discussion of the free will dilemmas.

finding the truth. In that respect it also deserved the name of *logica inventionis*, but in a new sense, namely that of the Cartesian rules of the search for the truth with the natural light of reason, which formed a certain linkage to the Platonic and Augustinian tradition with its key concept of illumination. The content of those rules was to be drawn from the experience of mathematics as that science which had to its credit the greatest cognitive success. That programme was pertinently formulated in the title of a textbook written by one of the leading Cartesians, E. W. von Tschirnhaus, published in 1667: *Medicina mentis sive artis inveniendi praecepta generalia.* The same Cartesian mainstream included the famous Arnould's and Nicole's Logic of Port Royal entitled *La logique ou l'art de penser*, first published in 1662, which only until 1736 had ten French and as many Latin impressions.

One could hardly deny the pertinence of Cartesian criticism when it was aimed at the triviality and sterility of such formal systems of rules as *pons asinorum.* In fact, an acute human mind never makes use of them in its search for the truth. Today we realize even better than the Cartesians did that inventive intuition is what can never be mechanized. Hence, should the only task of logic consist in the guidance or reinforcement of human minds, one could, and even had to, agree with the Cartesian critique of scholastic formalism. Yet it turned out, and that just in our century, that logic can enjoy other applications as well, being of fundamental import for civilization. It is the application that consists in mechanical processing of knowledge which is identified with the mechanical processing of texts, since only physical objects, which texts are, can be processed in a mechanical way.

That can be presented by reference to the bridge of asses. Let the memory of a computer hold the vocabulary of all those terms which can be used in syllogisms (here one sees reasons for finitism, as the set in question must be finite). Let there be given a grammar that makes it possible to produce, from already given terms, new and longer, but always finite ones. Let also the vocabulary include something which is used to some extent in the so-called thesauri that form a constitutive element of certain information languages (used in systems of information retrieval). That last element consists is a list of relations among the extensions of the terms involved, in particular the relations of inclusion and mutual exclusion, to which the rules of the bridge of asses refer. All this is, self-evidently, the same as the formulation of a certain axiom system of language, that system playing the role of axiomatic rules of sense as understood by Ajdukiewicz (1934). The axioms in that system should be independent of one another, which makes the system desirably economical. Endowed in this way the computer becomes a master of *ars inventionis* in the sense of scholastic formalists, for it can faultlessly find the middle term required in the proof of a given thesis. It will do that by searching the vocabulary and accepting for a given proof those terms which bear to one another extensional relations required by the rules of the bridge of asses. If it has to prove, for instance, that A is Z by selecting the appropriate middle term it will start searching the axioms in which those terms occur, and if it finds, for instance, the axioms $A \subseteq M$, $M \subseteq N$, $N \subseteq Z$, then it will give the answer in the affirmative with the indication of M and N as the sought middle terms – precisely in accordance with the first rule of the 'bridge of asses' as quoted above (the term 'follows' in it being interpreted in the sense of set-theoretical inclusion).

For our discussion it is of minor importance that the formal apparatus of scholastic logic is so limited as compared with the needs of the automatic processing of knowledge. Today we have at our disposal the apparatus of predicate logic with its various extensions which is incomparably richer than that of traditional syllogistic. The essential point, though, is to be seen in the similarity of programmes, which makes us understand the rules of logic as rules of operations on physical objects of a definite form (hence formalism) which at the point of departure constitute a discrete and finite set (finitism). That approach to logic, which may be read between the lines in Aristotle, was consciously taken by medieval formalists, and then developed by Leibniz and (independently of him) by later authors, especially George Boole; it was Boole likewise Leibniz from whom Frege took the idea of his logical enterprise (Frege 1880/81).[6]

[6]The mentioned year is the supposed date of finishing the manuscript which was successively offered by Frege to four mathematical journals but in each case was rejected by the editors as one which does not meet scientific standards. That manuscript is historically very interesting as it contains explicit references to Boole and Leibniz, and even a comparison of Leibniz's, Boole's, and Frege's own approach.

Thus the programme of formalization of reasoning, being a necessary prerequisite of its mechanization, has reached our times like the baton in a relay race and has become the contribution of that logical tradition to present-day problems of the mechanization of arguments.

2.3. The role of Lull and Lullism

2.3.1. Lullism was an important trend in philosophical thought for several centuries, beginning with the times of Lull himself and ending in the 18th century. In order to comprehend that importance, and also to perceive the misunderstanding, persevering to this day, when it comes to logic, we have to take into account the life story, personality, and legend of Lull.

Ramon (or, Raymundus) Lullus was born in 1232 in the capital of Majorca, an island recovered from the Arabs in 1229 by the Catalan army led by Jacobus I of Aragonia. Ramon's father, a rich burgher of Barcelona, served in that army. Ramon grew up in Majorca, where his father settled for good, in the court circle of the independent kingdom established after that conquest, and in the tradition of knights and troubadours. He soon made a career at the court and became the royal majordomo. That career was in a sense symbolic, because Lullus will owe many of his successes both during his lifetime and after his death to the support of the Aragonian royal house. He married in 1257 and soon became the father of two children, but this did not make him a stay-at-home; he remained by temper a troubadour and a warrior. He never renounced his family, but he was drifting more and more away from it until the moment when, at thirty, he had a vision of crucified Christ while he was writing a love letter to a lady. It was the moment of his conversion and the beginning of the great calling, which, it might be said, turned him into an intellectual missionary. He wanted to be martyred for faith when converting Muslims. Even though he lacked the requisite education he wanted to write a book that would be the best work in the world aimed at converting the infidels.

Being aware of the disproportion between his plans to become an excellent author and his even less than modest education he decided to go to Paris for studies, but following the advice of his prudent friends he decided to complete the first course of his studies in Majorca, which offered him the unique opportunity to become acquainted with the Arabic language and Arab science and philosophy. At this point we come to the problem that is fundamental for the comprehension of Lull's idea. Both Catalonia and Majorca were places where the Arab culture flourished. As is known, history destined that culture to become the great and creative intermediary between ancient thought and the European Middle Ages. That can probably be explained by the reach of the Arab empires which included the main centres of Hellenistic learning, and also by the flourish (not without the salutary influence of the Crusades) of commerce and banking, the factors which, as history shows, as a rule favoured cultural development.

That geographical region also witnessed the flourish of Judaic culture with its inclinations to subtle intellectual penetration and its mystique of number expressed in the doctrine of the cabala. That culture intertwined with both Arab and Christian thought. Thus the situation inspired one to a polemic dialogue and at the same time broadened one's horizon when it came to the methods of dialogue. The Arabs played the role of the pioneers of algorithmic methods (it was by coincidence, but deserving the status of symbol, that the terms "algorithm" and "algebra" were coined from Arabic words). Out of the works that emerged from the Judaic circles mention is due to the combinatorial interpretation of syllogistic in the Hebrew texts by a 13th century author named Albalag (discussed by Bocheński 1956).

When we see Lullus in such a historical context, at the crossroads of various cultural influences, we deprive him perhaps of the nimbus of a great originator and classic, but we perceive him as a remarkable link in that process of centuries from which also Leibniz drew his inspiration. Lull's subsequent life story, otherwise instructive and so picturesque as to have contributed to his later legend, need not be narrated here; what is most important for our story is to perceive that part of the process in question which is related to the mechanization of logic. We can already see how the idea of the algorithmic method, for which we have to be particulary grateful to the Arabs, was carried until the 17th century on the wings of Lull's legend.

2.3.2. Thus it only remains to mention some sources of that colourful legend. Lullus became a hero of two different and influential communities, the Franciscan and the Spanish. With the former he was connected not only intellectually, as a master of Neo-Platonic and Christian mysticism, cultivated by the Franciscans, but also organizationally as a Tertiary, i.e., a member of the so-called third order (for laymen) of the Franciscan community to which he was admitted in 1295. How his ideas were cultivated in the Spanish community can be seen, for instance, from the fact that the rulers of Aragon established centres of Lullistic studies. Both the Franciscans and the Spaniards contributed to his being worshipped as a saint, which was sanctioned by Pope Pius IX, who allowed his local cult in Majorca and in the Franciscan Order.

That *aura sanctitatis* was of considerable importance in the past centuries and was associated with that of missionary martyrdom. In fact, Lull – as we read in his autobiography which he dictated in Paris in 1311 – in his missionary travels to Tunisia and other Arab countries experienced things which could fill several novels of adventure, suffered imprisonment and persecutions, and twice barely escaped death.[7] Yet, in the light of recent studies, his death in Tunisia during his last travel in 1316 was not a consequence of persecutions; and so far we have to correct the popular version, repeated, among others, by Kotarbiński (1964). On the contrary, it was the time when Lull with quite a large group of his Catalan compatriots found himself in Tunisia in a privileged position because the sultan's throne had been occupied by an usurper who in his conflict with the opposition at home sought support on the part of the Catalans and even declared his readiness to become a Christian. It was precisely because of that declaration that the aged Lullus found himself with his mission in Tunisia again, this time as result of an agreement between that usurper and the king of Aragon.

Another element of Lull's legend is concerned with his alleged cabalistic writings, and hence work which enjoyed the aura of esoteric initiations and were much appreciated among the intellectual elite during the Renaissance. There is no reason to deny Lull's connection with the cabalistic doctrine because he had grown in its tradition, but we do not find in his work any direct references to the cabala, even such which mark the exposition of his logic by Giordano Bruno, and the cabalistic works ascribed to Lull have proved to be apocryphal.

[7]See (Bonner 1985), (Hillgarth 1971).

As will be seen below, also the belief that Lull invented logical calculus and the related mechanizability of arguments must be treated as a legend. It could have spread and win the admirers of Lull's *ars magna* either because of the ignorance of that art, which is comprehensible in the case of later authors, or because of a lack of a clear concept of logical calculus in the case of earlier authors.

2.3.3. Lull's famous invention discussed in the history of logic was called *ars magna* by himself. It covers a certain technique of forming judgements and also something which might be compared to an axiomatic system in the field of philosophy and theology. That system was conceived so that its axioms should belong to the common part of three doctrines, the Christian, the Moslem, and the Judaic, but with the special intention of a polemic dialogue with the teachings of Mohammed. The deductive apparatus was to derive, from those common assumptions, the tenets of the Christian faith, first of all those related to Trinity and Incorporation, and thus to refute the denial of those tenets found in the Moslem doctrine.

When it comes to the deductive apparatus it seems to the point to describe it with the words of T. Kotarbiński (1964), not only in order to make use of his pertinent formulations, but also in order to engage in a discussion with Kotarbiński's accompanying comment, typical of popular interpretations regarding Lull's role. Here is the text in question.[8]

> Lull strove to find a method of discussion and argumentation. That method was presented by him in his *Ars magna et ultima* and in *Ars brevis*, which was a summary of larger work. There are great wonders, there is plenty of theological fiction, but in the motley of vague naiveties we can find valuable methodological ideas. Thus Lullus takes as the point of departure a certain fundamental set of terms, among which he has names of attributes, names of relations, particles typical of questions of various kinds, etc. That 'alphabet' of the art is subjected by him to various orderings which make it possible to make a review of all possible wholes consisting of those elements, and finally he constructs a machinery consisting of concentric circles that move independently of one another. When those

[8]The text is an *ad hoc* translation from the Polish original (Kotarbiński 1957, p. 71).

> circles are turned one obtains, for instance, lists of triple of terms, and some of those triples may be terms of correct syllogisms. In this way one can find various predicates which are attributes of a given subject, or vice versa; one can also find the middle term for two given terms and the conclusion from given premisses.

And here is Kotarbiński's commentary to that summary of Lull's art.

> We thus have here like in a germ the later principle of singling out, in any reasoning, the totality of the objects taken into consideration, De Morgan's 'universe of discourse', and Descartes' rule of reviewing all possible cases, and finally, in a primitive form, Leibniz's later *calculus ratiocinator*, and even the nucleus of Jevons's machine for drawing conclusions.

Here are polemic comments to that account. The formulation 'like in a germ' seems to express a certain historical hypothesis, namely that the state referred to with these words is the first in a certain sequence of states or events. This is to mean even more, namely the relation of origin between that germinal state and the later stages of the process. However, contrary to Kotarbiński's interpretation, it can be hardly supposed that Descartes, a convinced and firm anti-Lullist, developed the nuclei found in Lull, the more so as Descartes had in mind mental steps in proofs and not a combinatorial mixing of terms. As to the universe of discourse in De Morgan's sense, it is a domain consisting of individuals, whereas Lull means an exhaustive list of abstract primitive terms.

Even if we construe that germinality in a broader sense, namely that of any origin, even a merely chronological one, i.e., without any genetic relation, his contention remains untenable. In this Chapter it has already been demonstrated that the most formal part of Lull's art, which might be termed combinatorial syllogistic, is to be found earlier in Albert the Great; the latter took it over from still earlier Arab authors, and the problem originated with Aristotle who asked about the way of finding the middle term. The fact that Lull's thoughts were imbued with that methodology can be sufficiently explained by his being versed in Arab logic, which he presented in his earlier writings, such as *Logica del Gazzali* (the Catalan version of the name of an Arab logician), written by him in Arabic and translated by himself into Latin and Catalan. Thus Lullus can hardly be regarded as the pioneer of the combinatorial and algorithmic interpretation of logic. It is another story that the belief of the later generations that he had been a forerunner in that field, combined with the admiration of his ideas and his person, played a significant role in propagating the programme of such logic.

2.3.4. There is still a semantic component of Lull's art, namely that metaphysical and theological doctrine which was to provide common assumptions for a discussion with the Moslem religion. We can have a good idea of that component from the description of discussion with the Saracens held by Lullus during his first visit in Tunisia. Here is Lull's report on his first missionary speech in Tunisia, translated into English by Bonner (1985, p. 34).[9]

> It is proper for every wise man to hold to that faith which attributes to the eternal God, in whom all wise men of the world believe, the greatest goodness, wisdom, virtue, truth, glory, perfection, etc., and all these things in the greatest equality and concordance. And most praiseworthy is that faith in God which places the greatest concordance or agreement between God, who is the highest and first cause, and His effect.
>
> However, as a result of what you have set before me, I see that all you Saracens who belong to the religion of Mohammed do not understand that in the above and other similar Divine dignities there are proper, intrinsic, and eternal acts, without which the dignities would be idle, and this from all eternity.

The core of this argumentation, in its most concise form, is to be found in the discussion which Lull had with an Arab dignitary during ones of his later missionary travels (the discussion ended in Lull's imprisonment at the moment when his Arab opponent ran short of arguments). Lull had earlier agreed with his opponent that their discussion would take as the starting point the common belief in goodness as one of the principal attributes (dignities) of God. In the speech quoted above the list of attributes was more comprehensive, and the full list is to be found in the exposition

[9] As for the fortunes of the Latin original see (Bonner 1985, p. 12ff); see also (Hillgarth 1971, App. X).

of the Art (for instance, in Ars brevis in the so-called figure A which lists the primitive concepts). The belief that God has all those attributes was part of the common heritage of both Christian and Arab metaphysics of those times, shaped in the Neo-Platonic melting pot. By departing from that common point Lull argued in his discussion, like in every other one in which he engaged, that in accordance with the Neo-Platonic principle (also common to his opponents) *bonum est diffusivum sui*, the goodness of God must spread of necessity. If that diffusion did not consist in giving rise to the second and the third person of the Trinity, then God would have to manifest himself only by the creation of the world, and then the manifestation of his goodness would depend on something which is not necessary, and hence imperfect and thus unworthy of God. In his first speech in Tunisia Lullus expressed that by saying that without that internal activeness within the Trinity God's *dignitates*, including his goodness, would be idle and hence would be not as perfect as it becomes the Absolute.

If one proceeds to read such texts written by Lull and describing the applications of his *ars magna* and takes into account the opinion of some authors that *ars magna* played an important role in the process of the mechanization of logic, then one must inevitably be surprised by the discrepancy between his texts and that opinion. It can be seen from those texts that Lull's main effort went in the direction of theological reflection; all that had only so much common with logic that he tried to find a heuristic aid. We do not find there any logical ideas that go beyond his contemporaneous canon (discussed earlier in this paper) of the combinatorial approach to syllogistic. Moreover, that tendency to formalize logic, typical of a certain trend in medieval logic, turns out to be weaker in Lull's case than, for instance, in the case of those who strived for algorithms like *pons asinorum*. This problem will be discussed in the sequel, concerning certain points of Lull's exposition of logic.

2.3.5. The exposition of *ars magna* (to be called hereafter the Art), in the version to be found in the opuscle *Ars brevis* (1308), begins with the Alphabet consisting of letters from B to K. Every letter takes on, according to the later context in which it is used, one of six meanings, and is thus marked by a certain systematic ambiguity (the letter A was reserved for the name of the first figure

– see below). In the first place on the list of the meanings of each letter Lull specifies the name of one of the attributes of God (*dignitates*). It is goodness under B, greatness under C, etc. In the second place we have the name of a relation (difference, agreement, opposition, etc.); in the third, an interrogative particle (whether, what, why, etc.); in the fourth, the name of an object (God, angel, man, etc.); in the fifth, the name of a virtue; in the sixth, the name of a moral vice.

The question immediately arises about the grounds for the choice of such and not other six categories (as the various places on each of the nine lists can be defined thus) and for the reasons of the choice of such and not other concepts within each category. Two reasons are plausible, and they perhaps complement one another: the existence of just such a conceptual paradigm in the intellectual circles which had shaped Lull's mentality, and the goal of the Art, which was the discussion of definite problems with the intention of converting Arabs to Christianity. The latter would mark the difference between the Art and the 17th century projects of a universal language based, as it was said, on a universal "alphabet of thoughts". In the case of those later authors, among whom Leibniz was the most eminent, some inspiration might have come not from Lull but rather from the legendary image of Lull and his Art, which fact, however, does not justify the ascribing that universalistic impetus to Lull himself.

The combinatorial procedure intended to serve the solutions of problems refers to the earlier distinction among three mental operations: simple apprehension, which results in a concept; formation of judgments; reasoning. The first operation yields that whole alphabet of concepts, grouped, as above, into six categories, each containing nine concepts. The formation of judgements has its analogue in the so-called three figures (i.e., a graphic presentation of possible combinations), and reasoning – in the fourth figure.

The formation of all possible judgements is carried out by turning two circles whose circumferences are divided into segments, each of them corresponding to one element of the conceptual alphabet. This is, obviously, based on the assumption, concordant with the aforementioned standpoint of traditional logic, that every judgement consists solely of two elements – the subject and the predicate (with the copula to link them together). There are as many as three circles which serve that purpose because one of them renders the combinations of attributes (goodness, etc.), the second deals with configurations of relations in the form of triangles (e.g., equal to, greater than, lesser than, each relation being symbolized by a different vertex of the triangle), and the third is obtained by the superposition of the first two circles, such that by moving the vertices of the triangles along the segments on the circumference of the first circle one obtains pairs of concepts (standing for judgements); one element of each pair is a relational concept, e.g., the judgement *goodness is greater* (thus what we call here relational terms by availing ourselves of modern terminology, in the grammar of the Art was treated as monadic predicates).

The fourth figure consists of three concentric circles, so that each triple yielded through turning these circles, for instance BDE, stands for two judgements, BD, e.g. *goodness is eternal* and DE, e.g. *what is eternal is powerful.* These substitutions of definite terms for various letters are made only by way of example because, as we know, every letter can be read in six ways, namely:
B: goodness, difference, whether, God, justice, greed;
D: eternity, opposition, out of what?, heaven, bravery, lust;
E: power, beginnings, why?, man, moderation, pride.

Hence, if we read the letters otherwise than previously while combining the segments in the same way, we obtain, among other things, the couple of sentences: *God is eternal* and *eternity has a beginning.* If the pairs of sentences obtained in this way are treated as premisses of syllogisms, then in the first case we obtain a true conclusion *goodness (in itself) is powerful*, and in the latter a false one: *God has a beginning.*

2.3.6. When a contemporary historian reads Lull's works in the light (or rather "in the darkness") of the opinions circulated about that author he will expect a mechanical method to decide which sentences are true and which reasonings are correct. He will, however, feel disappointed because nothing of that sort was Lull's intention. His real intentions must be interpreted in the context of the Neo-Platonic conception of cognition, assimilated by mystic trends in Arab thought, by Augustinianism, and also by Franciscan circles, that is the milieu that shaped Lull's thinking. This is how one should reconstruct the contribution of Neo-Platonism to the Art.

Simple apprehension, or that act of intellect which results in a concept, is not a construction but a recognition of what is *given*, either in the sense of Platonic anamnesis or that of Augustinian illumination. The very fact that something is "given" grants the truth to the judgment which is inherent like a germ in that idea or concept. For instance, when we have the concept of God given in this way, we arrive through that concept at the judgement that God has no beginning. What then is the purpose of the combinatorial Art so laboriously worked out? Its role is that of a perfect teacher, perfect because he is marked by infallibility which fallible human beings lack even if they have the dignity of teachers. It is the task of providing stimuli that release simple apprehensions and the resulting ideas (compare, among other things, Plato's dialogue *Menon* and St. Augustine's *De magistro*). Now the combinatorial properties of the Art guarantee that no possible stimulus, which here consists in combining concepts into pairs, will be disregarded. And to each such pair the human mind responds by itself by intellectually perceiving its congruence (as in the pair God–eternity) or incongruence (as the pair God–beginning). Thus Lull could have no reply to the question about the mechanical procedure of the verification of such truths, above all (regardless of all other reasons) because he did not face the problem of verification, and should anyone put such a problem before him, his Neo-Platonic formation would make him treat that question as a manifestation of ignorance.

The textual confirmation of such an interpretation is provided, among other things, by Lull's point in the last sentence of Part VII of *Ars brevis*; namely, he claims that the Art by guaranteeing the full review of all the possible combinations of concepts enables one to discover contradictions and impossibilities, and thus to debunk erroneous arguments of the sophists. That passage requires a comprehensive context covering not only written sources but even a certain folklore of mediaeval life (which has its echoes in witty sophisms, present even in conversations of common people, in Shakespeare's plays). Sophisms, consisting in the proving of things that according to the then prevailing opinions were impossible (those *impossibilia* which were referred to in the above-mentioned passage in Lull's *Ars brevis*), were a favourite pastime not only among scholars but also at courts and in inns.

Now, according to Lull, sophisms originate from incidental associations, and that results from the lack of systematic review of all possibilities. It is just the Art which prevents such a lack of completeness. Lull does not give examples, but we can exemplify his intentions through a sophism given by Siger of Brabant (who, like Lull, amply drew from Arab writings). It is said in it that no human deed is morally bad. Why? because to be bad it would have to be free, and it cannot be free since there holds that unquestionable principle that everything has its efficient cause and thus is fully determined.

Lull's list of concepts, as far as we know it, would not provide him with those needed to face the above argument; however, let it be extended with an imaginary example which might be in Lull's vein. Let us suppose that the list contains the theological concept of *concursus divinus* (much discussed in the 17th century). Then, when combining by means of Lull's circles the concept of human deed with every other concept, we obtain, next to the combination "human deed – (deterministic) cause", also the combination "human deed – concursus divinus", and since the divine concurring is to be interpreted as a non-deterministic cause, there is no reason to draw the conclusion that would deny the freedom of human deeds; this should discard the argument in question as an error of the *non sequitur* type. Even if the perception of the agreement between the concept of a human deed and that of divine concurring is the natural response of a theological mind, it may require a stimulus such as the recalling of the terms in question with a combinatorial survey. That can be done both by a good teacher and by Lull's "machine" as a teachers substitute. Lull regarded conceptual combinations made by the sophists as accidental and unnatural, whereas owing to the complete reviews ensured by the Art he hoped to reach those combinations which are fundamental and natural, their truth being grasped with "the natural light of the mind" (to use the Cartesian phrase, which is due to the same Augustinian tradition).

In a sense, the procedure like that may be interpreted as the search for a third term. In a debate, say, one denies that human acts are subjected to deterministic causation, hence are not free, and in order to prove his point one resorts to the concept of divine concurring as one that yields the middle term. Thus one argues: "Every act of human will is supported by divine concurring; every act supported by divine concurring is free, hence every act of human will is free." Note, however, that this kind of pursuing the third term is quite a different thing than that prescribed, e.g., for the *pons asinorum* procedure. No rules of the formal correctness of an argument are at stake but merely a heuristic device to activate memory, and so to find a concept which otherwise might have remained unnoticed. Once noticed, the concepts in question are perceived as connected in a necessary way, on the basis of an intellectual intuition. This should put an end to the legend of Lull as an author of a programme of arguments mechanization. Here are several places in *Ars brevis* to confirm our conclusion (numbers in parentheses refer to pages in (Bonner 1985)).

2.3.7. When describing the fourth figure, the one dealing with combinations of triples, Lull says that once such a combination is formed, the human intellect must grasp the *meanings* (italics by W.M.), i.e. some concepts, represented by the letters, and this is being done by seeking the agreement between terms and by attempts to avoid inconsistencies (p. 588); for instance, it would be inconsistent to combine terms "eternity" and "beginning" into the affirmative sentence "eternity has a beginning". Again, it is an intellectual intuition, and not a formal rule of inference, which should decide about the truth.

In Part VIII it is said that when a concept is to be employed, one should consider its definition. For instance, in order to derive a truth from the concept of divine goodness, one should take into account a suitable definition, e.g., that involving that God is eternal, to which Lull resorts to vindicate the existence of Divine Trinity according to the principle *bonum est diffusivum sui.* The argument is to the effect that God's goodness must have diffused itself through the whole eternity (not only since the creating of the world), and this could have been realized just by begeting the other divine persons. In order to show how the divine goodness produced various creatures, one should take into account a definition which in the case of, say, an angel will be different from those concerned with man, those concerned with a lion, etc. (p. 604; this seems to mean that God is good for men in a different way than for lions). There is, however, no algorithm of finding a proper definition which would yield the requested middle term; this relies on a philosophical intuition alone.

In Part IX, dealing with the properties of subjects of various kinds (God, angel, heaven, man, etc.), it is said about the angels that they can converse with one another without making use of the organs of speech and hearing; that is deduced from the definition of angel as a bodiless being and from his possessing the property "greater" as its attribute (that term is used as a monadic predicate). Namely, an angel is greater than a man, and hence man cannot be superior to him when it comes to the ability to communicate (p. 608). Again, this shows clearly that the strings of inferences typical of *ars magna* are produced by metaphysical insights and not by formal rules.

Even more telling examples are found in Part XI (p. 630), concerned with ways of answering questions, i.e. with the issue of problem-solving methods. For instance, by turning the three circles of the fourth figure we obtain the triple of concepts: B (goodness), C (greatness), D (eternity). This means a combinatorial generation of the problem: is goodness as great as eternity, i.e., is it infinite? (This is an echo of Aristotle's physics, in which eternity was, as it were, synonymous with infinity because of his rejection of all other actual infinities.) The answer is in the affirmative: it is argued that otherwise the greatness of eternity would not be good (here appears something like a vicious circle in arguing but Lull does not seem to have been too sensitive to such formal details).

In Part XIII, the last one, in which the method of using the Art is discussed, we find an advice for the teacher that he should make it plain to his disciples that the Art cannot be used without recourse to its close companions, that is, the reason, the subtlety of intellect, and good intentions (p. 645). In particular, the reference to good intentions, which is by far not incidental but results from Lull's mystical attitude, disproves the claim that Lull's Art, as a theory of argument, was a mediaeval anticipation of the programme of algorithmization or mechanization of proofs. Even if there were such anticipations of formalistic approach to reasoning in Lull's times, they are to be sought (as shown above) in the nominalist, i.e., terminist trends, and not in his intuitionistic doctrine.

* * *

Was it necessary in this chapter to dwell so much on Lull's story to show how irrelevant it is to the process of the mechanization of reasoning? The question is not uninspiring. Its examining should bring deeper insights into the process discussed.

A promising way to gain them starts from reading Leibniz's dissertation *De arte combinatoria* (see the full title in References) in which Lull's project appears in a very wide spectre of similar enterprises, all of the grown in the same, so to speak, culture of syllogistic; in that culture the reduction of logic to certain subject-predicate forms created the illusion of the decidability of all problems through exhaustive combinations of subjects and predicates.

Thus the said title proves symptomatic of that philosophy of knowledge, and of the resulting research programmes, which started with Aristotle, and did not entirely disappear until the scientific revolutions of our own century (quanta, relativity, evolutionistic cosmology, Gödel's limitative theorems). When taking advantage of the cosmological concept of stationary universe (as opposed to the evolving universe), and linking it with the familiar logical idea of the universe of discourse (as the totality of the things under study i.e., those represented by variables), we can introduce the analogical pair of notions: that of the *stationary universe of discourse* and that of the *evolving universe of discourse.*

The combinatorial approach is associated with the vision of the stationary universe of discourse, in which the set of concepts forming our knowledge, and thus constituting the universe of discourse, is definitely fixed and closed. Then the whole cognitive enterprise would consist in searching for those combinations of members of the universe which yield true propositions, while logician's task would involve providing people with a proper method of doing that. It was just the programme of the famous *logica inventionis* as mentioned in the subtitle of *De arte combinatoria.* The combinatorial approach, though, turned out quite sterile as far as the development of knowledge is concerned, since the universe of discourse of scientific theories rapidly expands as ever new concepts appear, and that would require ever new recombinations.

On the other hand, the combinatory strategy is fairly suitable in the problem-solving as carried out by computers. It can be seen in the mechanical checking of a proof, where each formula has to be resolved into its atomic components being combinatorily surveyed in order to detect inconsistencies among them, if there happen to be any. Besides that combinatorial feature, the mechanization of reasoning requires its, so to speak, physicalization in the sense that abstract entities, as truth-values, consequence operations, etc. should be represented by some physical things in order to be menageable by a physical mechanism. As discussed earlier in this chapter, the latter aspect of mechanization should be traced back to the medieval nominalists who conceived logic in that "physicalist" manner, i.e., as dealing with visible forms of expressions, while Lullus was far from being such a formalist.

However, as to the combinatorial approach to reasoning, Lull should be merited for its dissemination in the centuries who were to follow, as his ideas were extensively commented by his successors. Leibniz reports on Lull's project among many other ones being discussed in his *De arte combinatoria*, and soberly criticizes it for its deficiencies; the author whom Leibniz praised most is Thomas Hobbes to whom he owed the idea that *every* mental operation consists in computing (*Hobbes merito posuit omne opus mentis nostrae esse computationem*, Sec. 63).

How should one estimate the weight to be attached to either of the components of that process which prepared the mechanization of reasoning, the combinatorial component and the formalistic one? According to the present author it is the latter which is by far more significant. In this perspective, Lull's absence in that process, the absence documented by texts revealing his intuitionistic attitude, is a fact discrediting his legend. Nevertheless, the combinatorial aspect should not be entirely disregarded. It played a crucial role in the erroneous belief in the stationary universe of discourse, i.e., that human knowledge can be definitely and safely established by combinatorial procedures. That human brains were mistaken for computers in those ages in which nobody dreamt of the latter, it was one of those errors which paved the way to new understandings.

CHAPTER THREE

LEIBNIZ'S IDEA OF MECHANICAL REASONING AT THE HISTORICAL BACKGROUND

3.1. An interaction between logic and mathematics

3.1.1. The historical process which eventually resulted in the creation of computing and reasoning machines should be conceived as one of those great steps which were decisive for the evolution of mankind.[1] Among the steps of such consequence there was the emergence of language in a remote past, the birth of the idea of number, and certainly the rise of logic in some writings of Plato and Aristotle.

Since Aristotle up to our times, and specially in the 17th century, logic was addressed to human beings to improve their intellectual performances. Those intentions notwithstanding, the most important final result of that process, which started with the rise of logic, consists in the invention of reasoning machines. That human minds can do without a logical theory, when relying on their inborn logical capabilities alone, is obvious when one observes the history of discoveries and other manifestations of creative thinking. But the mechanical mind's inferences are due to some devices for which a logical theory forms an indispensable foundation.

As claimed in the preceding Chapter, the possibility of mechanical reasoning depends on the existence of what we call the *logical form* of a linguistic expression, that is to say, a structure determined by those terms which are relevant to logical validity. When

[1]This chapter partly coincides with a part of (Marciszewski 1994), viz. the chapter entitled 'Formalized versus Intuitive Arguments: the Historical Background'. In that book the rhetorical point of view implies seeing logic as the theory of a mental mechanism, most relevant to producing arguments. Since such a mechanism can be studied through computer simulations and through interactions with computers, there is a vital point common to the books in question. Hence the same historical background is discussed in both of them.

a reasoning is so formulated that its validity can be recognized on the basis of its form alone, we call it a *formalized reasoning.* It may be called mechanized as well, but in a broader sense, namely that a mechanical procedure, i.e., one that does not appeal to our understanding of the content, is sufficient to judge the validity.

However, when we take the term 'mechanized' in a strict meaning (as is the case in this volume), then formalization is a preparatory step towards *mechanization.* The latter consists, so to speak, in expressing the logical form in a language of physical states subjected to some causal laws of physics, be it mechanics (as, e.g., in Babbage's machine), be it the domain of electronics (as in modern machines). Thus the logical operations is identified with physical processes occurring in a machine, and this, let us repeat, yields the mechanization of reasonings in the strict meaning. This explaining of the role of formalization of reasonings as crucial for their mechanization should make obvious why so great import is attached to the former in the story being told in this volume.

The merit of founding logic goes back to Plato and Aristotle, the former as the one who discovered logical validity, the latter as the discoverer of how that validity depends on logical form. As for Plato's contribution, the core of Socratic method of argument, as presented in Plato's dialogues, consists in discerning between the truth of a statement and its being deduced with a valid inference (even in our times, after so many centuries of the development of logic, people happen to have difficulties with that distinction). In a typical argumental dialogue, Socrates helps his interlocutor to deduce some consequences from a view which the partner claims to be true. When, nevertheless, the consequence proves false, the partner is bound to acknowledge the falsity of his initial view; this reveals how the validity of an argument can be independent of the truth of its premisses. Such is the constant strategy of Socrates, though no attempt is made to theoretically justify that practice.

To justify it, one should have resorted to the form of the sentences involved in an argument. This was done by Aristotle by introducing letters to represent variable contents (subjects and predicates) while the form, rendered by expressions as 'every ... is', 'is not', etc. remains constant. Though originally there was no direct translation of that form into physical states of a machine, the later development from syllogistic to Boole's algebra of classes, and then the transforming of the latter into Frege's algebra of truth-functions, and next the invention of the method (due to C. Shannon) to represent truth-functions by some states of an electrical device, has led to the nowadays mechanization of reasoning.

This line of development can be highlighted by the negative example of Lull's method as discussed in Chapter 2. In spite of its appearance of mechanical (because of those rotating wheels) Lull's art was no step toward mechanization of arguments, for it did not involve any thought of mathematical operations. It was Hobbes who perceived an analogy between reasoning and computing, and then it was Leibniz who enthusiastically endorsed that idea and contributed to its accomplishment in an algebraic manner. This is why Leibniz so appreciated the idea of logical form as found in some medieval students of syllogistic, as akin to the form of algebraic calculations. Let the following quotation render that Leibniz's attitude.[2]

> The invention of syllogism is one of the most important and finest inventions of the human mind, and it is that kind of mathematics whose significance is not sufficiently known; and it may be said that it contains the art of infallibility if we only know how and are able to make use of it.

Leibniz pointed to the mathematical nature of logic, which was possible in his times owing to the advances in algebra. Before we discuss that connection in a greater detail, let it be noted that there was another relation between logic and mathematics which stimulated the development of logic in a broader sense, i.e., as including methodology of deductive sciences; in that development the deductive method was brilliantly exemplified by geometry. Historians of Greek science pronounce the names of these two disciplines in one breath when they, for instance, refer to *the climate of logic and geometry, which we have known so well since Euclid* and *the impetus imparted to Greek geometry and logic* by the mathematical methods of the Babylonians (De Solla Price 1961).

The methodological feedback between logic and geometry consisted in the fact that the ideas worked out in logical theory used to pass to the praxis of geometricians, who in turn provided logicians with the best patterns of proofs. Aristotle, when writing

[2]See Book IV, 17, of *Nouveaux essais sur l'entendement humain.*

his *Analytics*, drew from the interpretations of geometry that were known to him the pattern of necessary knowledge and reliable inference, while Euclid, when writing his *Elements* half a century later, availed himself of the methodological concept of *common axioms* ($\tau\alpha$ $\kappa o\iota\nu\alpha$), that is, those which are not specific, say, to geometry, but are drawn from some more general theory; note in this connection that the Aristotelian example of such an axiom — "if equals be taken from equals the remainders are equal" — occurs on the list of axioms in the first book of Euclid's *Elements*.

3.1.2. The second encounter between logic and mathematics was due to *algebra* and took place for the first time in the 17th century. Algebra itself as the science of the solution of equations, that is finding the unknowns in equations, originated in Babylonia in the third millennium B.C., and found special application in the computation of shares in inheritance cases. Shortly afterwards it came to be known in Egypt, which is testified by papyri dating from ca. 1800 B.C. When arithmetic and algebra, born in Babylonia, met in the Hellenistic period with Greek geometry and logic, they gave rise to that gigantic 'tree of knowledge' of which our civilization is the fruit (cf. de Solla Price 1961). Arab scholars made great achievements in the development of algebra, continued by European scholars. This applies in particular to Al-Kwarizmi (9th century A.D.), who also contributed to the spreading of Hindu numerals, later called Arabic numerals. The very word 'algebra' originated from the title of his work that begins with the words 'al gabr' (the whole title meant 'the transfer of a constituent from one side of an equation to the other').

The decisive advances in algebra were made possible by the development of symbolic language, which was the contribution of the Europeans from the 13th century onwards (the Arabs, like the Babylonians before them, used in their computations phrases drawn from natural language). Symbols of arithmetical operations were coming successively into use, and the breakthrough was marked by the introduction of letters as symbols of numbers (F. Viète, 1591); thus the idea of a *variable* entered mathematics. The realization of that breakthrough manifested itself in the birth of the term *analytica speciosa* to single out the mathematics using such notation that one symbol does not correspond to a single object (e.g., the figure

'2' to the number 2), but to a class, or *species*, of some numbers (the Latin adjective *speciosa* means also 'beautiful', 'magnificent', etc., hence it may have also reflected the fascination with that invention). It was Descartes who became the coryphaeus of that analytics when in 1637 he published his *Geometry*, to which *Discours de la méthode* was the annex; in this work he gave a synthesis of geometry with algebra or analytics (hence the term 'analytic geometry').

Owing to the maximal generality which the notation using letters gave to algebra, Leibniz was in a position to realize that a letter need not refer to a class of numbers, but can refer to any class of objects of any kind. The use of letters was invented for logic already by Aristotle, but only the successes of algebra could give birth to the idea of an algebraic treatment. That was one of the greatest of Leibniz's ideas concerned with logic. It was partly materialized by himself, but it was first published in print two hundred years later, after the same discovery had been made in the meantime by other authors. Yet the permanent imprint of algebra upon the mentality of people living in the 17th and the 18th century consisted in their experience of how an appropriate language renders thinking more efficient and signally contributes to the solution of problems (cf. Chapter 4). This fact played its role in the development of a movement for improving the whole language of science.

3.1.3. The second discovery of the algebra of logic took place in England in the mid-19th century, and was due to a Pleiad of prominent algebraicians, of whom G. Boole (1815–1864) rendered the greatest services to logic.[3] He was helped in that respect by the then nascent comprehension of the abstract nature of algebra, which is to say that an algebraic theory does not refer to any specified domain (which was particularly emphasized by G. Peacock (1830). Algebra can, on the contrary, find application or, more precisely, *interpretation*, in a class of structurally similar domains that can be described with the means provided by a given algebraic theory. In this way algebra, after having developed from the old

[3]The rise and development of the algebra of logic in the 19th century is comprehensively dealt with in Chapter 5 of this volume. However, for the readers' convenience some basic data are given in this Chapter to discuss earlier stages in the light of the further development.

science of solving equations, has become the most general theory of structures, that is systems of objects for which certain operations are defined. Such operations are described from the point of view of their properties, e.g., commutativity or its lack, or the performability of operations with a neutral element (such as zero for addition in a certain algebra having an interpretation in the arithmetic of natural numbers), and the like. One of the interpretations of this calculus called *Boolean algebra* (since Boole was its main promoter) corresponds to that part of logic which is known as the truth-functional calculus. The same algebra has another interpretation in traditional syllogistic, and it was just that interpretation which was so penetratingly anticipated by Leibniz.

In order to fully grasp the meaning of that interpretation we have to refer to Boolean algebra in its contemporary form. It has at least two operations (the others can be introduced by definitions), one of them called complement and symbolized by '$\sim$', and the other called multiplication and symbolized by '$\circ$'. Moreover two objects of those operations are considered; they are singled out from the set of all those objects with which a given theory may be concerned, and are denoted by the symbols '1' and '0' (other notations are also used, but we choose the one which is most convenient typographically). Multiplication satisfies the conditions of commutativity and associativity, and such as: $X \circ 0 = 0$, $X \circ 1 = X$, while complement is defined by the conditions: $\sim 1 = 0$, $\sim 0 = 1$.

Boolean algebra can be interpreted arithmetically, the domain of natural numbers being limited to zero and unity (in such a case all the laws of that algebra are satisfied). It can also be interpreted in many other ways, but two interpretations, already mentioned above, one in the domain of sentences, and the other in that of sets, are fundamental for logic.

In the domain of sentences, multiplication is interpreted as the linking of sentences by 'and' (conjunction); complement is interpreted as the negation of a given sentence, 1 as truth, and 0 as falsehood. If a formula consisting of those symbols and variable symbols standing for sentences is always true, that is, if it is true regardless of whether truth or falsehood is assigned to the variables which occur in that formula, then it is a law of logic and belongs to the logical theory termed truth-functional, or sentential, calculus. By way of example let us mention the law of contradiction, which reflects the truth of the idea that it is not so that something is and is not the case, in symbols: $\neg(p \wedge \neg p) = 1$ for all substitutions, that is for both $p = 1$ and $p = 0$. By making use of the symbols of conjunction ($\wedge$) and negation ($\neg$) we can express all the laws of the sentential calculus, but in order to bring it near to the reasonings found in science and everyday life it is convenient to introduce other connectives which as a rule occur in reasoning, such as 'if p, then q', which we define by the formula $\neg(p \wedge \neg q)$. It states that, in accordance with our understanding of the truth of the conditional sentence, it is not so that the antecedent holds (i.e., is true) while its consequent does not hold. We can likewise define the sentential structure 'p or q' by using the formula $\neg(\neg p \wedge \neg q)$.

3.1.4. When interpreting Boolean algebra in the domain of sets (which traditionally were referred to as extensions of names) we can in a natural way express the four traditional kinds of categorical sentences, which form the building material of syllogisms. The result of the operation of multiplication, i.e., $A \cdot B$, is this time interpreted as the common part of the two sets, that is as the set of those objects which are in both A and B. The complement of the set A, i.e, $-A$, is interpreted as the set consisting of all objects (in the domain under consideration) which are not in A. Further, the symbol '1' now means the set of all objects in a given domain, and '0', the empty set, i.e., the set which has no element. And here are the interpretations of the categorical sentences in Boolean algebra.

Every A is B: $A \cdot -B = 0$; for instance, that every lion (A) is a predatory animal (B) means the same as that the set of those lions which are not predatory animals is empty, which is to say that there are no such lions; this condition is accompanied by the reservation (made already by Aristotle) that the set A is not empty, which is to say that $A \neq 0$.

No A is B: $A \cdot B = 0$; e.g., the fact that no lion is a hare means the same as that there are no lions which are hares (the reservation of the nonemptiness of A is valid as above).

Some A is B (in other formulations: Some A are B; There is at least one A which is B): $A \cdot B \neq 0$; e.g., the fact that a certain lion is tame means the same as that the set of tame lions is not empty, i.e., that there are elements of our universe of discourse (for instance, the domain of animals) which are lions and at the same time are tame.

Some A is not B (other formulations, as above): $A \cdot -B \neq 0$; e.g. the fact that a certain lion is not tame means the same as that the set of lions which are not tame is not empty, in other words, that there are lions which are not tame.

Such a translation of traditional logic into Boolean algebra enables us to activate a strong deductive apparatus of algebra for obtaining economical and elegant proofs of the theorems of traditional logic. Such was also the intention of Boole (1847) himself, which can be seen in the title of his work: *The Mathematical Analysis of Logic, Being an Essay toward a Calculus of Deductive Reasoning.*

The anticipation of that interpretation, to be found, in Leibniz's treatise of 1686 (first published in 1903) entitled *Generales inquisitiones de analysi notionum et veritatum* (inquiry into a general theory of concepts and judgements), differs from the modern form mainly by the fact that in place of the symbols '$\neq 0$' it has the Latin phrase 'est ens', i.e., 'is an entity' (or, briefly, 'est', i.e., 'exists'), while the symbols '$= 0$' have the analogue in 'non est ens', i.e., 'is not an entity' (or 'non est', 'does not exist'). These phrases can be interpreted in at least two ways: as stating the existence of objects which are the *extension of a certain concept*, i.e., a certain set (extensional interpretation), or as stating the existence or non-existence of combinations of properties, i.e. the *intension of a certain concept* (intensional interpretation). Leibniz himself was open to both interpretations; he valued the extensional one as technically efficient, but believed the intensional one to be better because of certain philosophical considerations. The idea of the translation of a sentence consisting of subject, copula, and predicate, such as "every man is intelligent" into an existential sentence of the kind "non-intelligent manhood does not exist", was of scholastic provenance, to which Leibniz clearly referred. The brilliance of his own idea consisted in noticing an analogy between such sentences and equations of the algebra he had formulated (which included the operation of linking concepts and negating a concept). In those equations on the one side we find such a combination of concepts (one of which may be negated), and on the other, one of the two 'magnitudes' expressed by the terms 'ens' and 'non-ens'.

The above stating how algebra came to link logic to mathematics finely illustrates two fundamental laws in the history of human thought, one of which we may name (in accordance with Leibniz) the law of continuity (*lex continui*) and the other the law of the logic of development. Both of them motivate certain rules of historical research, which are as significant as the rule of the study of the past from the position of the present. Continuity consists in the fact that historical reality does not make leaps (*natura non facit saltus*, as Leibniz used to say), even though that property of the process of history is hardly perceptible in view of our irresistible, and otherwise legitimate, inclination to divide human history into periods (which suggests demarcation lines between epochs), and also in view of the recently fashionable tendency to stress revolutions in the history of science. And yet ideas move forward by making steps so small that they can be treated as being close to zero. The impression of leaps which we may be have due to defects in our knowledge (caused by gaps in the sources or by insufficient research) and also to abbreviations and simplifications that are inevitable in view of the boundlessness of the continuum of history. That law of continuity results in the research directive stating that for every process that shows gaps we have to seek sources and facts which would fill those gaps with intermediary links. Guided by this directive, we should see the algebra of Leibniz's logic in connection with the uninterrupted development of algebra that was going on for some time (especially when it comes to algebraic notation), and also to the continuation by Leibniz of the ideas of scholastic logic, a continuation which consisted above all in grafting them onto the tree of algebra.

The case of the algebra of logic also illustrates the second of the regularities mentioned above, namely the law of the inner logic of the development of thought. Owing to a historical accident, Leibniz's texts were not published early enough to influence the development of the algebra of logic, but even though that accident could have delayed the process of development, it could not stop it for good. That was so because the very content of traditional logic included the nuclei of its further development toward the computational approach (e.g., the use of variables, and the fact that the relation between the subject and the predicate lent itself to the extensional interpretation). And if an idea contains an important element, then that fact has such a gravitational influence upon human minds that finally someone notices that element and articulates the consequences inherent in it. This again leads us to the methodological conclusion that in reconstructing the lines of development of a given idea, we have to follow them from the inner content of that idea which will lead us to the appropriate sources, and on obtaining new source data we have to compare them with the results of earlier deduction. That idea guided the present volume inscribes in the genealogical tree of logic not only Leibniz and Boole, but also the numerous algebraicians living in the 16th, 17th, 18th and 19th centuries, and even those Schoolmen who toiled to express sentences with the subject-and-predicate structure in the form of sentences on existence or non-existence.

3.1.5. What lesson can we draw from the above story of the algebra of logic? Note that the historian's task is to trace the paths of thought in order to distinguish the highways from side roads, to watch turns and forks, and to find those paths which lead us farthest into the future. But to trace the direction in which human thought moves one must have — as in the case of tracing the movement of a body in space — at least two points. By having one such point in the 17th century, and after having found the second in the 19th century, one could state that the destiny of logic led it toward the *algebraic interpretation*. That observation gave rise to new questions: How far could logic have advanced by going in that direction? Did that direction determine the future of logic for good, or did it merely define one of the stages? If the latter was true, what was the next stage to be, and why did it take that form rather than another? We have to answer these questions briefly before we proceed to discuss the other factors of development which had their sources in the 17th century. The founders of the algebra of logic, that is G. W. Leibniz and G. Boole, and also A. De Morgan (1806–1871), E. Schröder (1841–1902), C. S. Peirce (1839–1914), and others, were convinced that the whole of logic can be contained in algebraic calculus. They were also aware of the fact that traditional logic, even after its algebraic reconstruction, lacked the means required to describe relations (for instance, it was impossible to render in its language even such a simple relational sentence as "for every number there is a number greater than it"). That was why the next stage was to consist in adding the algebra of relations to the already existing algebra of sets (i.e., the analogue of the traditional theory). These new researches, carried out mainly by De Morgan, Schröder, and Peirce, gave rise to the theory of relations, which became an important and indispensable discipline in the border area between logic and set theory, but its language (combined with that of the algebra of sets) did not suffice to express mathematics in its entirety, either.

At this point one could ask the question: Why should the language of logic perform such an important function? The intention of 19th century algebraicians was merely to improve traditional logic from the mathematical point of view. The reply is that the late 19th century, regardless of the intentions of the various authors, witnessed the need, and at the same time the possibility, of constructing a universal symbolic language of mathematics in which well defined symbols and precise syntactic rules would replace the vocabulary and syntax drawn from natural language. A great step forward in that direction was made in the period from the 15th to the 17th century, which saw the spreading of the symbols of arithmetical operations and relations. But at least one more step remained to be made, namely the rendering of the logical form of sentences by means of an appropriate symbolism and related specialized logical syntax.

Such a step was vigorously planned in the 17th century, when the idea was advanced of constructing a rigorous and universal language of science, ideographical in character, called *characteristica rationis* (reasoned or conceptual writing), *characteristica universalis* etc. But at that time linguistic and logical means (and perhaps also the appropriate philosophical approach) were not available. On the other hand, those algebraicians who were active in the mid-19th century, even though they worked out some such means, neither planned such a great work nor were in a position to do it, because that required going beyond both Aristotelian logic and the algebraic schema.

At the turn of the 19th century logic entered a new path by providing mathematics with a universal and rigorous symbolic language, which had also far-reaching (though not universal) applications in other disciplines and in everyday discourse. For that to happen the 17th century programme of a universal language had to be revived, which did take place owing to the publication for the first time, in 1840 (by J. E. Erdmann) of some logical writings of Leibniz which formulated that programme (e.g., *De scientia universali seu calculo philosophico, De natura et usu scientiae generalis, Fundamenta calculi ratiocinatoria, Non inelegans specimen demonstrandi in abstractis*). There were three founders of contemporary logic, independent of one another when it comes to ideas even though they had intensive contacts with one another: Gottlob Frege (1848–1925), Giuseppe Peano (1858–1932), and Bertrand Russell (1872–1970). All of them referred to Leibniz's ideas, and two of them believed themselves outright to be the executors of his testament by carrying out his programme of universal language. One of them was Frege, who undoubtedly has first claim to this title. In 1879 he published a work which presented the whole of contemporary logic (i.e., the sentential calculus and the predicate calculus) under the title *Begriffsschrift*, meaning conceptual writing, and can in Latin be pointedly rendered by Leibniz's term 'characteristica rationis'. His reference to Leibniz's terminology was intentional, which is proved by the fact that two years later he wrote a paper entitled *Booles rechnende Logik und die Begriffsschrift* (which remained unpublished after having been rejected by editors of three leading periodicals), where he explicitly referred to his intention to carry out Leibniz's programme.[4]

3.1.6. We have discussed so far three approaches of logic to mathematics, each of them bringing new links between the two. The contact with geometry consisted in the fact that a similar conception of deductive science was propagated by logical theory (Aristotle's *Posterior Analytics*), and put into practice by geometry. Either trend somehow influenced the other from the 4th century B.C. until the 17th century A.D. and beyond. The influence of logic upon geometry found a strong reflection in the oft-recurrent problem of the logical independence of Euclid's fifth axiom from the remaining ones, which reached an epoch-making peak in the birth of non-Euclidean geometries.

The 17th century saw a new connection because, on the one hand, variable symbols came to be consciously used in algebra, which in logic had been a spontaneous practice since Aristotle, and on the other, the similarity which was borne out in this way promoted the algebraic interpretation of logic, achieved by Leibniz.

---

[4] A comprehensive examination of Leibniz's influence on Frege is found in Kluge's writings (1977), (1980), (1980a).

Even though Leibniz's endeavours at first fell into oblivion, the fate of the idea itself proved independent from historical accidents, and the idea materialized again, first in attempts of some mathematicians in the 18th century (cf. Chapter 4), and then, most successfully, in the works of Boole and other 19th century algebraicians (cf. Chapter 5).

But not all of Leibniz's ideas were revived in the work of those later algebraicians, because the notion of a universal language of science, in which the algebraic calculus would be built in as an instrument of deduction, remained as it were hidden in shadow. It was brought to daylight again only by Frege and Peano, congenial readers of Leibniz's texts, but with the essential correction that the idea of a universal language was to materialize not in the form of an algebraic calculus but in the form of the much more powerful predicate calculus.

The whole story has been narrated so far on two temporal planes, that of the 17th century and that of our times, with the additional incursion, made in a perspectivic shortening, into antiquity, in accordance with the principle of explaining the past by the present and also by its own past. In moving further in that direction we have to mention one more, the most recent, rapprochement between logic and mathematics. This new connection is totally outside the perspective of the 17th century, which is why it should also be mentioned here, in order to show not only the proximity of that epoch relative to ours but also the distance between the two.

When new logical calculi developed at the turn of the 19th century, namely the predicate calculus (which overcame the limitations of algebra) and the sentential calculus (which may be treated as a certain interpretation of Boolean algebra), new vistas were opened to the logicians. On the one hand, it was now possible to develop and improve the calculi themselves by formulating ever new versions and by constructing other calculi that could be superstructured upon the former (such as modal logics). On the other, since those calculi abounded in problems to be investigated, it was also possible to pose questions about the consistency of each of them, their completeness, the purposes they could serve, the relations among the axioms and concepts of a given system (the problem of independence), and finally the relations among the various systems (the interpretability of one of them in terms of another, the consistency of one of them on the assumption of the consistency of another, etc.). Precisely the same questions can be posed about mathematical theories. Among these, the arithmetic of natural numbers is of particular importance from the logical point of view (it was first axiomatized by Peano, and superstructured on logic and set theory by Frege and *Principia Mathematica*), because the remaining branches of mathematics can in a way be reduced to arithmetic: the 19th century witnessed the reduction to the latter of the arithmetic of real numbers, to which geometry had been reduced earlier owing to Descartes.

Thus, when at the world congress of mathematicians in 1900 D. Hilbert presented to the mathematical community the task of proving the consistency of mathematics (in view of the antinomies discovered at that time), it was known that it would suffice to concentrate on the problem of the consistency of the arithmetic of natural numbers using for that purpose the logical apparatus, both that which was already known at that time and that which was still to be created. We also owe to Hilbert the term 'metamathematics' to denote such studies. They were carried on intensively in the 1920's and 1930's, and brought astonishing results due to Hilbert's companions as well as critics, such as J. Herbrand, G. Gentzen, A. Turing, K. Gödel, A. Tarski.

As metamathematics developed it was more and more penetrated by mathematical concepts and methods as indispensable instruments of research. They were in particular drawn from arithmetic (e.g., the use of recursive functions initiated by Gödel), algebra (e.g., Tarski's algebraic approach to non-classical logics), topology and set theory (which from its very inception was connected with logic). We also have to record the interpenetration of logic with mathematical linguistics and automata theory. In this way metamathematics, which was originated as the logical theory of mathematics, itself underwent mathematization.

3.2. The Renaissance reformism and intuitionism in logic

3.2.1. The development of all ideas, and hence also that of logic in the 17th century, is immersed in the melting pot of general civilizational development, in the cultural ferment of the period. The present Section is to deal with that problem. The century which will be described here brings out with particular clarity the fact that the masterpieces of intellect grow from the roots of tradition but also from a lively dialogue with the milieu that is contemporaneous with their authors. And that milieu means not only congenial individuals, but also the audience consisting of readers, disciples, opponents, snobs, etc., in a word, the entire enlightened public. What that public believes, to what it aspires, what concepts it uses to grasp the real world, what it reads and what it discusses – all these are factors of vital significance for intellectual enquiries.

Moreover, one has to bear in mind the relationships between science and the totality of culture, on the one hand, and the economic conditions of a given society, on the other. The flourishing of the arts and sciences coincides as a rule with a flourishing economy, as is indicated by the history of Miletus, the golden age of Athens, the Italian cities in the late Middle Ages, the Netherlands, England and France in the 17th century, Germany in the 19th century. Hence if the ordinary man participates in the creation of economic well-being, then he contributes to the creation of spiritual values as well. It is worth while in this connection to quote the following description of the epoch now under consideration:[5]

> Then, in the late sixteenth and early seventeenth centuries, had come the wave of commercial and financial expansion – companies, colonies, capitalism in textiles, capitalism in mining, capitalism in finance – on the crest of which the English commercial classes, in Calvin's day still held in leading-strings by conservative statesmen, had climbed to a position of dignity and affluence.

Economic expansion favoured advances in science among other things by the demand for the services of mathematicians: Simon Stevin (1548–1620) worked out a notation for decimal fractions, intended to facilitate the computation of bank interest in that capital of finance which the Netherlands were at that time for Europe. Pascal constructed a calculating machine for the needs of his father who was a tax collector. The same expansion gave rise to large numbers of skilled technicians who were necessary to advance science. Leibniz, for instance, when leaving for Paris in 1672, expected that he would find a mechanic there who would help him

[5] See (Tawney 1937, p. 182).

construct the prototype of a calculating machine. The demarcation line between scientists and craftsmen was liquid at that time, which is illustrated by the large number of craftsmen in the London Royal Society, the greatest collective authority in science in the 17th century. Another interesting confirmation of that fact can be found in the correspondence between Leibniz and the young Christian Wolff. When the latter offered Leibniz his dissertation on cog wheels, Leibniz in his kind reply suggested that Wolff should consult German craftsman on that matter. In this way a blacksmith or a locksmith as potential partner in the discussion with Wolff participated in the formation of scientific culture of their times. One more fact from that period is worth mentioning in this connection: the deliberate coining of scientific terms in national languages so that navigators and clockmakers, not versed in Latin, could study mathematics, astronomy, etc. It were the Dutch who excelled in that: to this day they have terms dating from that time which in no way resemble their Latin equivalents. The advances in mechanical and optic crafts, in printing and in other arts and their impact upon the advances in science (especially in the construction of research instruments) is a topic which deserves lengthier discussion, but here it is enough to recommend it to the reader's attention.

Let us look now at the other direction in that relationship: how the spiritual climate of the period shaped the development of finances, commerce, travels, and inventions. The spirit of expansion had its religious motivation in the Puritan doctrine as well as a secular motivation in the ideology of 'the kingdom of man', propagated as early as in the 15th century.[6] The finest advocates of the latter included Francis Bacon (1564–1626) and also William Shakespeare (1561–1626) as the author of the dramatic fairy tale *The Tempest,* in which the wizard Prospero proves able to control the elements. In Book I of aphorisms, concerned with the explication of nature and the kingdom of man, included in *Novum Organum*, Bacon wrote emphatically that human reason must be fully freed and purified so that the same road should lead to the kingdom of man, based on sciences, as that which leads to the kingdom of Heaven, which no-one can enter unless one becomes like a little child (i.e., free from prejudices like a child; aphorism 68). The mastery of Nature, lost as a result of the original sin, is to be regained by technical

[6]See (Tawney 1937), (Swieżawski 1974, p. 149).

inventions: no authority, no sect, no star has influenced humanity more than mechanical inventions have (aphorism 129). The author of these words was not only a well-known philosopher but also the Chancellor of the Kingdom of England, and hence a London craftsman could feel fully appreciated and encouraged to develop his ingenuity, the more so as from the stage of Shakespeare's theatre The Globe he could hear flattering ideas e.g., that an inventor resembles a wizard who controls Nature.

There was one more heritage of the Renaissance from which the 17th century benefited, namely Pythagorean and Platonic philosophy, from its earliest beginnings most closely intertwined with mathematics (what is believed to have been announced at the entrance to Plato's Academy). The Platonic trend was always present in European thought, especially from the time when it was reinforced by the ideas of Plotinus (204–270) and in that new form, called Neo-Platonism by historians, established contacts with young and vigorous Christendom, in both its orthodox version and that marked by the influence of gnosis. A significant example of that synthesis can be seen in the person of Proclos of Constantinople (410–485), one of the widest known commentators of Euclid and a Platonizing theologian, to whose ideas Kepler used to refer. The first Christian writers and Church fathers were as a rule Neo-Platonians. This applies also to the greatest of them, St. Augustine of Hippo (354–430), whose ideas were truly reincarnated in the 17th century in the work of Descartes, Pascal, and the milieu of Port Royal, and in the late 19th century came to the rescue of G. Cantor, the author of the theory of infinite sets, in his clash with the philosophy of mathematics that followed the Aristotelian approach to infinity.

The Pythagorean and Platonic trend particularly penetrated the enlightened public from the 15th century, which witnessed the renaissance of Platonism in Italy, especially in Florence, until the 17th century, when that trend reached Cambridge, and in that way influenced Newton. The spread of Platonism at that time can be traced in two ways. One can follow its more popular version, manifested in the interest in the mysticism of numbers, the cabala, astrology, and the accompanying emergence of societies concerned with those matters. Another form of that popular version can be seen in the presence of certain Platonic elements in the teaching of mathematics as the methodological pattern of all science.

The other path on which we can trace the reach of an intellectual current is the number of outstanding authors who endorse it. Here are some prominent personalities of that time connected with the propagation of the Platonic vision of the world: Marsilio Ficino (1433–1499), Nicolaus of Cusa (1401–1464), Leonardo da Vinci (1452–1519), Nicolaus Copernicus (1473–1543), Galileo Galilei (1564–1642), Johannes Kepler (1571–1630). This wave of Platonism had its share in preparing the next one, which came in the 17th century. The latter less frequently referred to Plato himself, but it was nevertheless influenced by Platonic views on the possibility of and need for the mathematization of all knowledge and on the *a priori* nature of cognition.

Thus, there are in the 17th century two vast currents, one of them linked to technology and economics and ideologically based on the catchword 'the kingdom of man', and the other permeated by Platonic metaphysical speculation. Both influenced logic when postulated that logic should give the human mind an instrument whereby it could arrive at the truth. While they had one and the same goal in view they disagreed as to the method of reaching the truth: the former staked on induction, whereas the latter, preoccupied with mathematics in the Platonic manner, postulated a purely deductive method (*more geometrico*). Subsections 3.2.2 and 3.2.3 present the former and the latter, respectively.

3.2.2. The advances in logic in the 17th century, the authority which that discipline enjoyed and the hopes it had aroused originated from the faith in the power of reason, and also the faith that this power can be increased. The expansion of logic was to be achieved by its development and popularization after the appropriate reform which it had to undergo: the logic inherited from Aristotle and the Middle Ages, the advocates of the reform claimed, would not prove equal to the task.

Before we tell the story of the reformers we should note, in order to obtain a proper background, that the teaching of logic was marked by the conservative trend observable in some university towns (Helmstedt, Wittenberg, Giessen, Königsberg). That trend developed in the 16th century mainly owing to Philip Melanchton (1497–1560); being associated with this co-founder of Protestantism it was called a variety of Protestant logic. (There was also a reformist trend, which originated from Peter Ramus, who enjoyed great authority among the Protestants as the martyr who lost his life on St. Bartholomaeus Night.) The conservative trend reverted to Aristotle, and hence to the source, in the spirit of the Renaissance and the Reformation, thus disregarding the novelties contributed by scholastic logic. Though the conservative approach did not dominate in the 17th century logic, it should be mentioned as a party to the polemics and as the contrastive background against which one more clearly sees the reformist claims.[7]

Another contrastive background must be seen in the preceding period. The idea of a reform of logic developed in the 16th century; the 15th century was still busy with commenting the classics, namely *Summulae logicales* of Petrus Hispanus (1220–1277: as Pope he was known as John XXI), and later achievements, such as those dating from the 14th century. A special merit for that continuation went to Paul of Venice (d. 1429), whose *Logica Parva* (1428) continued to be taught at many universities for two centuries to come. A certain logical idea of Paul of Venice in his *Logica Magna* found its way to the writings of Leibniz and came to play a certain role in the development of extensional logic.[8]

The history of the reform of logic begins with Peter Ramus (Pierre de la Ramée, 1515–1572), remembered as a vehement opponent of Aristotle whom he blamed for being unnatural in his approach to logic. Nevertheless, he took over a great deal from Aristotle, namely the theory of definitions, the theory of judgement, and the theory of inference. Two things made him differ essentially from Aristotle, but one of them was in the sphere of programmes (in which one can relatively easily become an innovator), and the other has its source in age-old tradition, but not the Aristotelian.

In his programme Ramus included the requirement that logic be adjusted to natural human thinking, without that artificial abstractedness for which he blamed Aristotle. His followers took up those slogans, and for the two centuries that followed we find in the title of numerous logical compendia various psychological terms

[7]A detailed factographical discussion of logic in the period under consideration is given by Arndt (1965).

[8]See (Nuchelmans 1983, Sec. 11. 3. 1).

connected with the conception of logic as the science of living human thinking and not of any abstract entities.

The best known titles of this kind include the following: *La logique ou l'art de penser* by A. Arnould and P. Nicole (first published in 1662, it had had ten French and as many Latin impressions by 1736); *Medicina mentis sive artis inveniendi praecepta generalia* by E. W. von Tschirnhaus, 1687 (several other courses in logic had the same title presenting logic as the way of healing one's mind); *Introductio ad philosophiam aulicam, seu ... libri de prudentia cogitandi and ratiocinandi* by Ch. Thomasius, 1688 ('philosophia aulica' meant court philosophy, which required — as claimed the second part of the title — prudence in thinking and reasoning). The same may be said about the title of the work written by Ch. Wolff (1679–1754), a philosopher who at first was an adherent of the Cartesian school in logic, and hence an opponent of syllogistic, but later, under Leibniz's influence, became an advocate of syllogistic. He was also the principal commentator of Leibniz's philosophy in Germany. And here is the characteristic title of his book: *Vernünftige Gedanken von den Kräften des menschlichen Verstandes und ihrem richtigen Gebrauche in Erkenntnis der Wahrheit*, 1713 (Reasonable Thoughts about the Powers of Human. Understanding and their Proper Use in the Cognition of the Truth); the book had had 14 German impressions by 1754, and also four impressions of a Latin version and two impressions of a French version. Even though in its content Wolff's work did not belong to the psychological trend, it was assimilated to it by its title, which shows the strength of the trend represented by Ramus and terminated only in the late 19th century by the concentrated attack against *psychologism in logic* from the position of the mathematically-oriented philosophy of logic (G. Frege, E. Husserl, J. Łukasiewicz, and others).

Another novelty relative to Aristotle's ideas consisted in the division of logic into the science of judgements and inference and the science of making inventions, in which much space was dedicated by Ramus to definitions (the connection between the method of making inventions and the formulation of definitions will be fully seen in Leibniz). It was not an absolute novelty, because such a division was introduced by the *Stoics* more than a dozen centuries earlier and was transmitted to the Middle Ages by Boethius (480–524), but turning it into the principal claim which fitted with the aspirations of the epoch was the work of Ramus.

When it comes to Francis Bacon, not only one half of logic but the whole of it was to consist, in his opinion, in the art of making discoveries which would pave the way for the kingdom of man. Bacon designed a logic of *induction*, which was totally to replace the existing logic of *deduction*. The characteristic feature of deduction is that the conclusion does not contribute any information that would be not contained in the premisses: it cannot convey more information than the premisses do. He was not the only one to attack logic in this way, because a whole chorus of critics, among whom the voice of Descartes sounded loudest, blamed syllogistic (the only theory of deduction universally accepted at that time) for not contributing anything new to our knowledge. But while Descartes and his followers wanted to replace syllogism by another deduction, modelled on the experience of mathematical thinking, whose creative character would not be questionable, Bacon, guided by an empiricist ideology, failed to perceive such an alternative solution. Hence, if logic was directly to serve the expansion of human knowledge, he had to stake everything on the inductive method. But at that infantile stage of the logic of induction people failed to realize that for an *increase of information* one has to pay with a decrease of certainty, or, to put it more precisely, a lessening of probability which has certainty as its upper limit. This is so because a general law is intended to be a conclusion drawn from observations to infinitely many possible cases, and therefore tells us more than the observations reflected in the premisses, which always cover finitely many cases. But it is uncertainty which is the price paid for such an increment of knowledge, because while Nature may to a given moment confirm that general conclusion, there are no reasons to be sure that it will confirm it in each subsequent case.

If this is so, then the logic of induction is not any alternative solution to the logic of deduction but can at most be its complement (which, by the way, to this day is still in a stage of planning, rather than in that of achievement). Bacon's ideas, even though they finely expressed the spirit of the epoch, did not pave any new ways for logic despite the intention of their author (that found reflection even in the title of his work, *Novum Organum*, which referred to

Organon, the title of Aristotle's logical writings). The greatness which logic was destined to attain and which it did attain in our century, was reached by the old Aristotelian road when it merged with the path along which mathematics was developing. Those two paths came closer to one another for the first time in the 17th century, and the process was due to Leibniz. Thus it was he and not Bacon who became the forerunner of future logic. And those thinkers who, while underestimating Aristotle, did not commit the mistake of underestimating deduction, also proved to have come closer to what was ahead. That process will be discussed in what follows.

3.2.3. The problem of how logic should be is not merely a matter of logical theory. It deeply penetrates the philosophical vision of the world and the mind. Bacon's design for a reform of logic originated from the empiricist conception of cognition which ascribes to the human mind the role of a passive receiver: it is like a screen on which Nature casts the image of itself through the objective of man's sense organs, and if the mind is active in any respect then only by posing questions to Nature.

Such empiricism is opposed by apriorism, also termed rationalism (in one of the meanings of that word), which states that there are truths which man comes to know before all sensory experience; they are given to him as it were in advance (Latin *a priori*). In that sense, reason (Latin *ratio*) is independent of experience. According to the rationalists, the essence of cognition consists in the realization by the human mind of those truths, given it a priori, which are general and necessary. Premises drawn from sensory experience and experimental data may be helpful in that process of deduction, but they never suffice to lay the foundations of our knowledge. Hence logic as an instrument of cognition, interpreted in the rationalist sense, must be a logic of deduction which not only serves the presentation and ordering of the truths that are already known (in the 17th century, scholastic logic was blamed for being confined to that role), but it can also lead one to discoveries and reveal truths that had been unknown earlier. This is best proved by the role of deductive reasoning in mathematics; at the same time, mathematics offers irrefutable examples of the existence of a priori truths.

The fertility of deduction in the sphere of mathematics fascinated outstanding minds in the 17th century, for it was imposing as compared with the sterile, it was believed, pedantry that marked the theory and praxis of deduction in its scholastic version. That was why the new programme of deductive logic (the adjective 'deductive' will henceforth be dropped) postulated the study of those methods of thinking which had been intuitively used by mathematicians of all times; those methods were to be formulated in plain und clear rules and would thus yield a truly useful logic. It was on that path, it was expected, that the art of creative thinking consisting in arriving at new truths (*ars inveniendi*) would be propagated.

Before we examine the implementation of that programme we have to pay attention to the peculiar difficulty that is inherent in it. Deduction consists in transferring to the conclusion the content, that is information, already contained in the premises, hence in the conclusion there can be as much information as there is in the premises but no more. On the contrary, a discovery consists in arriving at new information one did not have previously. This is why (deductive) logic as an instrument for making discoveries may seem unthinkable. But whoever raised such an objection against the concept of creative deduction would be guided by a conception of the mind that differed from that which marked the 17th century reformers of logic. The opinion characteristic of those times had its roots in Platonism, which dominated enlightened minds in the 16th and the 17th century from Florence to Cambridge (two famous strongholds of Platonism). In the Platonic interpretation, apriorism involves a certain conception of what we would today call subconsciousness. In Plato himself that conception was associated with the theory of anamnesis, which meant cognition as reminiscence of truth originating from the pre-existential phase of the mind. Hence a discovery meant nothing else than bringing to daylight a truth that was formerly concealed. In other words, it could metaphorically be described as the shaping of that truth as it were from a substance which was given earlier, in the similar way as a bud and a fruit are shaped from the substance of a given plant.

That theory was linked to, and brought in fuller relief by, the characteristically Platonic conception of the teacher as someone who helps his disciples to bring to light what was in the shade or a penumbra. Socrates as described in Platonic dialogues defined that activity by metaphorically comparing it to obstetric art, and St. Augustine coined the term *illuminatio* to name that process whereby human thought comes to light. This gave rise to the conception of dialogue as illuminating interactions, and also the conception of dialectic as the theory and art of dialogue which has deduction as its main instrument (which can clearly be seen in Plato's dialogues). Dialectic later came to be termed logic. Thus from the very inception of Platonism logic was to serve the process of causing the truths that are innate to the mind to pass from the state of latency to that of clear presence. Thus it is not new contents that we owe to logical deduction but a new way in which they exist in human minds. And the fundamental question of the new programme of logic was as to what rules are to be used in decoding the truths encoded in our minds.

3.2.4. The classical answer to this question was given by René Descartes (1596–1650), followed by a group of authors collectively called the Cartesian school. It included Baruch Spinoza (1632–1677) with an opuscle *Tractatus de intellectus emendatione* (on the improvement of one's mind), E. W. von Tschirnhaus (1651–1708) as the author of *Medicina mentis* (see 2.2 above), and also the authors of the Logic of Port Royal, whose popularity greatly contributed to the spreading of the Cartesian methodology.

Descartes's famous method is expounded in his two works whose very titles express the idea of the art of making inventions, ars inveniendi. One of them is the *Discours de la méthode pour bien conduire sa raison et chercher les vérités dans les sciences*, published in Leyden in 1637 (the climate in the Netherlands was more favourable to new ideas than the vicinity of the Sorbonne). That good guidance of one's mind, mentioned in the title of the *Discours*, was presented in even greater detail in the treatise *Regulae ad directionem ingenii* (rules for the guidance of one's mind), published posthumously in Amsterdam in 1701. The latter work offers a more detailed insight into the Cartesian method (it presents 21 rules while the *Discours* only four), but the former was the text which continued to influence the community of scholars throughout the first half of the 17th century, and this is why we shall concentrate on it. Here are the rules which tell the readers how to guide one's mind to arrive at the truth. The first of them states the famous principle of Cartesian doubt. Let these rules be distinguished by numbers, each number preceded by 'DM' for *Discours de la méthode.*

> DM-1. Not to accept anything as true before it comes to be known as such with all self-evidence; this is to say, to avoid haste and prejudice and not to cover by one's judgement anything except for that which presents itself to the mind so clearly and explicitly that there is no reason to doubt it.
> DM-2. To divide every problem under consideration into as many particles as possible and is required for its better solution.
> DM-3. To guide one's thought from the simplest and easiest objects in order to rise, later on, slowly to the cognition of more complex ones; in doing so one has to assume regular connections even among those which do not form a natural series.
> DM-4. To make everywhere precise specifications and general reviews so as to be sure that nothing has been omitted.

Characteristically enough, the *Discours* in its first edition appeared not as a separate item but as the annex to the opus whose role in the history of mathematics and philosophy is comparable to that of Euclid's *Elements*, namely *Géometrie.* Analytic geometry, expounded there for the first time, combined two branches of mathematics which had previously been developed separately, namely geometry and the arithmetic of real numbers with accompanying algebra. That work used the concept of function in an advanced way and essentially contributed to the invention by Leibniz and by Newton (independently of one another) of mathematical analysis (differential and integral calculus). That gigantic step toward the integration of mathematics (the next comparable one was only the origin of set theory in the late 19th century) aroused such an admiration and such hopes in the contemporaries that the second half of the 17th century was animated by the vision of a future science, termed *mathesis universalis*, which would cover the whole of knowledge, philosophy included, in the form of a single mathematized theory.

What was the advantage of the combination in one publication of an exposition of geometry and reflection on method? Now the exposition of geometry itself could be used, from the methodological point of view, as the model for the deduction of theorems from initial self-evident facts formulated in axioms, but it did not provide any advice upon how one is to arrive at the self-evident statements needed as premises of proofs. That required a description of the appropriate operations performed by the human mind, and just such a description was provided by the *Discours de la méthode.*

The problem of finding self-evident premises needed to prove theorems expanded in the Cartesian school and its milieu to a vast set of problems pertaining to the distinction between the deduction of consequences from theorems already known, which was then called the *synthetic method*, and the search for premises to support a given theorem, called the *analytic method.* Here is the statement of that distinction in the Logic of Port Royal, whose authors refer in a note to a manuscript by Descartes, to which they were given access.

> Method can be described generally as the art of a good arrangement of a series of thoughts in arguments used either for the discovery of truth, if it is not known, or for its demonstration to others, if it is already known. There are thus two methods. One of them is the method of discovering the truth, termed analysis or the method of solution; it might also be termed the method of invention. The other is the method of making the truth which has already been found accessible to others. It is termed synthesis or the method of composition, and it might also be termed the method of exposition. (Book IX, Chap. 2).

The terms *method of solution* and *method of composition* which occur in that description were to render a similar distinction existing already in scholastic logic and earlier in Euclid and Pappus.

The theory of the analytic method, that is of the *ars inveniendi*, on which the efforts of that generation of philosophers were focussed, evolved in two opposite directions, the psychological and the formalistic. The psychological trend, which more and more concentrated on advices pertaining to the attitudes and behaviour of the mind (with the accompanying moralizing tendency) was particularly strongly voiced in the writings of Christian Thomasius (1666–1728) and in those of his adherents, grouped in the then recently (1649) founded university in Halle. Thomasius, who followed

Ramus in that respect, shared with the latter the (so to speak) practical orientation in logic. His disciples and adherents were strongly influenced by Protestant pietism, which at that time had its centre in Halle, and were thus influenced by the sentimentalist movement with its inclinations to irrationalism. No wonder, therefore, that ultimately the idea of the methodological identity of mathematics and philosophy, and hence the idea of *mathesis universalis*, was abandoned by them. This ends, in the late 17th century, one of the ramifications in the evolution of logic.

But the 17th century also witnessed the emergence of an antipsychological trend, which announced the birth of mathematical logic and made first steps in that direction. That trend, in distinction from the former, will here be termed formalistic, even though that name is not quite suitable. That trend, too, raised the problem of the art of discovery, but freed it from psychologism, owing to the idea of algorithm. Those plans and ideas, even though they did not emerge from a historical vacuum and, as usual, marked the fruition of the work of numerous generations, are — in historical transmission — associated with the name of Leibniz.

3.3. Leibniz on the mechanization of arguments

3.3.1. When we discuss the past logic in the light of what is logic today, this perspective does not predetermine the choice of the instruments of interpretation. From the sets of concepts which we have at our disposal we shall select that which includes the concept of *algorithm* as a recipe for the mechanical performance of tasks of a given kind; this kind of concepts should best help us to see the future of logic in its civilizational context. Before we proceed to discuss that concept (see Subsec. 3.2 below) we have to concern ourselves with that of logical form as the one which played an essential role in the development of logical algorithms.

The concept of logical form has developed from the observation that certain sentences, e.g., in English, are accepted as true solely on the basis of their structure, with disregard of empirical and any other extralinguistic data. Such are, for instance, the following sentences.

It rains or it is not the case that it rains.

If it thunders, then it thunders.

It is not the case that it (both) dawns and does not dawn.

If it dawns, then the cock crows, and hence if the cock does not crow, then it does not dawn.

If a certain shoemaker is a lover of music, then a certain lover of music is a shoemaker.

$2 = 2$.

The law is the law.

It is said about such sentences that they are accepted solely on the strength of their logical form, and they are accordingly termed logical truths. In order to define the concept of logical truth in a general manner, and hence also the concept of logical form, one makes use of the partition all of expressions of a given language into two classes: one of them includes the expressions called *logical terms*, such as 'and', 'or', 'not', 'if ... then ...', 'some', 'every', '=', and the like, while the other includes all the remaining expressions, called *extralogical.* Owing to this distinction, the intuition inherent in the formulation that logical truths are accepted solely on the strength of their structure, can be made more precise in the following way. A sentence is called a *logical truth* if it remains true after the replacement in it of extralogical expressions by any expressions, drawn from the same grammatical category — with the proviso that such a replacement is made consistently, which is to say that on replacing A by B in one place we do so in all those (and only those) places in which A occurs in that sentence. Thus the sentence 'It rains or it is not true that it rains' is a logical truth because we obtain a true sentence if we replace in it the sentence 'it rains' by any other sentence, for instance 'The world was created in six days'.

Such replaceability is of great significance for the recording of logical truths. Since for their being true it is quite indifferent what extralogical expressions occur in them, it is both possible and very useful to replace these or other extralogical expressions by letters which are like the blanks to be filled. Those letters should at the same time indicate the grammatical category of expression which may be used to fill the place occupied by a given letter in the schema. This is why special conventions are adopted in that matter; for instance, the letters 'x', 'y', 'z' indicate the place of proper names, the letters 'p', 'q', 'r', the places of sentences, etc. In this way we obtain schemata of sentences which are always true, for instance the following.

p or it is not true that p.

If q, then q.

$x = x$.

If p, then q; hence, if not-q, then not-p (where 'not' is synonymous with 'it is not true that').

Logical truths expressed in such a schematic form are called *laws of logic*.

The fact that logical truths are truths solely in virtue of their form paves the way to the construction of algorithms of two kinds: those which give us a mechanical method of deciding whether a sentence is a logical truth, and those which give us a mechanical method of deciding whether a sequence of sentences claimed to be a so-called formalized proof is in fact a proof of that kind. The invention of algorithms which would serve those purposes was one of the great plans of Leibniz who expected that the invention would result in an epoch-making progress of human knowledge. His expectations came true only in our century, and that was accompanied by the discovery of the fact that they can come true only to a certain extent, because it is not possible to algorithmize the whole of mathematics (Gödel's famous result proving the incompleteness of arithmetic), and hence that it is *a fortiori* impossible to algorithmize the whole of our knowledge.

But even those partial solutions are important enough to treat the concept of logical form (which paved the way for them) as one of the greatest attainments of the human mind. Here are several data from the history of that discovery, or rather the process of discovering, because here, too, we deal with a process which continued for centuries. The first flash of the intuition concerned with logical form must be seen in indirect reasoning, that is reasoning by reduction to absurdity, which is to be found in the enquiries of early Greek mathematicians, and which is particularly frequent in Plato's dialogues (where as a rule it is used by Socrates, who treated it as main tool of his dialectic). An *indirect proof* starts from the assumption that the statement to be proved, say C (for conclusion) is false. If this assumption results in a contradiction, this means that it is false; but if the assumption of the falsity of C is false, then C itself has to be true.

The very fact of making use of such proofs points to the sensing of logical form, because in an indirect proof two desirable properties of our thinking are separated clearly from one another as if in a prism: the factual correctness, that is truth, of the premisses and conclusions, and the formal correctness, that is agreement of inference with the laws of logic.

Aristotle of Stagira (384–322 B.C.), to whom goes the unquestionable credit of creating the first system of formal logic, often made skillful use of indirect proofs, but what made him the founder of logic was the fact that in formulating the laws of inference he used for that purpose letters in the role of name variables. This is not to say that he already had a clearly shaped concept of logical form, but it is beyond doubt that his epoch-making invention in the sphere of notation guided the thoughts of all his followers toward the idea of that form.

Logicians from the philosophical school of the *Stoics* (active for five centuries, beginning with the 5th century B.C.) were undoubtedly aware of the formal nature of logic obtained as a result of the use of variable symbols. They used numbers as sentential variables and thus created the nucleus of the logical theory which we now call the *sentential calculus*, while Aristotle's *syllogistic* was a *calculus of names.*

The wealth of ancient logic, including the teaching of the Stoics, was transmitted to later generations by Boethius, a Roman statesman from the times of Theodoricus. Logicians active in the late Middle Ages made a successive important step forward by linking the logical form to the functioning of those expressions which were termed *syncategorematic* and analysed by them in detail. The realization of the fact that syncategorematic expressions determine the logical form of a sentence is particularly clear in the case of John Buridan (rector of Paris university in the first half of the 14th century). We find in Buridan a conception of logical form which resembles the contemporary one and includes the distinction between *categorematic terms* and syncategorematic ones, the latter being those which determine the logical form of a given sentence. He also took into consideration the impact of the categorematic terms upon the logical form; for instance, the form of a sentence 'A is A' differs from the form 'A is B'. Such a concept of logical form was taken over from Schoolmen by Leibniz, who turned it into the focal point of his conception of logic. Yet he remained isolated in that respect, perhaps because of the strong antiformalistic and antischolastic tendencies which emanated from the milieu of both Ramus and Descartes. That concept was rediscovered only in the mid-19th century by Augustus De Morgan, one of the founders of the algebra of logic.[9] Thus Leibniz had to play the historical role of being the last Schoolman and at the same time the first logician of our epoch.

3.3.2. Cartesian methodology called for a direct contact of the mind with the object of cognition. Leibniz's methodology pointed to the indispensability of another way of thinking, in which the contact of the mind with the object of cognition is not direct, but takes place through the intermediary of signs as those instruments of thinking which represent the object assigned to them. One of the ways of using signs as instruments consists in our grasping with our thought the sign instead of the object (its designatum).

That intermediate way of thinking about things, in which the thing itself is not present in the sphere of our attention, was called blind thinking (*caeca cogitatio*) by Leibniz. Operations on large numbers are its simple example. As long as we remain in the sphere of small numbers, for instance when multiplying three by two, we still can be guided by some image of the object itself; e.g., we imagine an arbitrary but fixed triple of things and join to it one triple more. Such an operation can be performed physically or mentally even if we do not have at our disposal symbols of a numerical system. It is otherwise when we have to multiply numbers of a dozen or so digits each. In such a case we are deprived of that visual contact with objects and have to rely on sequences of figures which represent them. Such sequences are physical objects assigned to abstract objects, namely numbers, and operations on

[9] See (Bocheński 1956, Sec. 42.01).

figures are unambiguously assigned to the corresponding operations on numbers. For instance, juxtaposition, that is writing one after another the symbols '2', '·', '3', is an operation on signs, and the corresponding operation on numbers consists in multiplying two by three.

If we add to this description of blind thinking one element more, then we obtain the concept of algorithm, which played the key role in Leibniz's logic. The term 'algorithm' itself was not used by him in its present-day meaning. That role was played by the term *calculus*, i.e., computation, or by metaphorical expressions like *caeca cogitatio* and *filum cogitationis* (thread of thinking). The latter term referred to Ariadna's thread, that is to the tool which enabled Thesaeus to move about in the Labyrinth and to find his way out, that is to solve his problem without any thinking, and without the knowledge of the Labyrinth itself. The description of the algorithm is to be completed by the statement that it is a procedure consisting of a series of steps such that every step unambiguously determines the next one in accordance with precisely defined rules. Those rules determine the successive steps in the procedure under consideration by referring solely to physical properties (such as size and shape) of objects, e.g., the shape of figures, and the positions they occupy. One can carry out the algorithm of addition in columns without having the slightest idea of numbers and availing oneself only of the rule which states that if the sign '2' is written under '3', then one has to write '5' in the next row in the same column, etc., and of the rule that one has to start such operations from the rightmost column and after filling it to pass to the next column on the left until the very end (that is to say, until one reaches the leftmost symbol, which is repeated in the last row if there are empty places above or under it). On joining the self-evident condition that the number of the steps must be finite we obtain the full description of algorithm, which can be summed up as follows.

An *algorithm* is a recipe for a procedure pertaining to a definite class of tasks (e.g., addition of sequences of figures), which in each case guarantees the obtaining of the correct result after the performance of a finite number of operations, and that owing to the fact that the objects of such operations (e.g., sequences of figures) are reliably identifiable owing to their perceivable physical properties (e.g., shape, position).

When the idea of algorithm came to be mentally worked out by Leibniz, it had had a picturesque past which is a good example of the harmonious work of the history and the cooperation of various civilizations. The idea and the term are associated with Al-Khwarizmi, the Arabian mathematician in the 9th century, who systematized computational methods in his manual by means of sequences of instructions. When those methods became popularized at first in Italy (Leonardo Pisano, 1180–1240) and later in Germany (Adam Riese, 1492–1559) and in other countries, people started speaking about 'computations in Al-Khwarizmi's sense', that is, in accordance with such detailed instructions. Al-Khwarizmi's manual was based on a Hindu astronomical work dating from the 7th century A.D., which a certain Hindu brought to Baghdad in 773. It included the decimal system, invented in India ca. A.D. 400, which offered an opportunity for such mechanical computations as those referred to above (which could be subsumed under 'blind thinking').

3.3.3. In that melting pot of ideas which was Leibniz's mind, the idea of mechanical computations merged with his looking at logical form through the prism of algebra as it was known at that time, and also with two other ideas, which he had received from the past but transformed and combined in his own way. One of them, propagated especially by Joachim Jungius of Hamburg (1587–1657), consisted in the opinion that the essence of thinking consists in the analysis of concepts, that is in decomposing compound concepts into simpler ones, until one arrives at the simplest ones possible. Those simple and non-decomposable elements form, as Leibniz used to say, the alphabet of human thoughts. He saw the model of such a procedure in the decomposition of a number into factors which yields a product of prime numbers.

Another idea, very popular in the 17th century, was that of a universal ideographic language, whose single symbols would be assigned not to sounds but, as in Chinese, directly to concepts. That planned language was then called philosophical language, conceptual script, universal script, etc. Here are, by way of example, two titles of works on that subject that were popular in that time: Gregor Delgarno published in London in 1661 his treatise *Ars signorum vulgo character universalis et lingua philosophica* (The art of signs or universal script with a philosophical language), and then in 1668, also in London, *An Essay Toward a Real Character and a Philosophical Language* was published by John Wilkins.

According to Leibniz, this conceptual language should be constructed in compliance with Jungius's idea, which is to say that simple symbols, elements of the alphabet, should represent the simplest non-decomposable concepts, while combinations of symbols, analogically to algebraic operations, should express appropriate compound concepts. The rules pertaining to such operations on symbols were to form a logical calculus that would be algorithmic in nature and would guarantee error-free reasoning:

> *Ut errare ne possimus quidam si velimus, et ut Veritas quasi picta, velut Machine ope in charta expressa deprehendatur.* (So that one could not err even if one wished to, and would find the truth given as it were *ad oculos*, rendered in writing by the machine).

This formulation, to be found in Leibniz's letter to Oldenburg (dated October 28, 1675), looks like a prophetic description of the computer which, containing in itself an algorithm of proving theorems or of verifying proofs carried out by man, produces the result on the screen or on a sheet of paper coming out from the printer. It is also said in that letter that the truth would be delivered in a visible manner and irrefutable owing to an appropriate mechanism (*mechanica ratione*). His similar idea was that one could easily avoid errors in reasoning if theorems were given in a physically tangible manner (*si tradentur modo quodam palpabili*), so that one could reason as easily as one counts (*ut non sit difficilius ratiocinari quam numerare*).

The substantiation of that programme is to be found in Leibniz's letter to Tschirnhaus (May 1678):

> *Nihil enim aliud est Calculus, quam operatio per characteres, quae non solum in quantitate, sed et in omni ratiocinatione locum habet.* (Computation is nothing else than operation on graphic symbols, which takes place not only in counting but in all reasoning as well).

Such computation was sometimes called by Leibniz *calculus Ratiocinator*. On other occasions he used the term *characteristica Rationis*, as was the case in his letter to Rödechen of 1708:

> *Characteristicam quandam Rationis cujus ope veritates rationis velut calculo quodam, ut in Arithmetica Algebraque, ita in omni alia materia quatenus ratiocinationi subjecta est, consequi licet.* (A conceptual symbolism, by means of which one could arrive at truths of reason as it were by computation, like in arithmetic and in algebra, and in every other matter if it can be subjected to reasoning.)

Such a strong call for the computationalization of any domain of human thinking resembles the contemporary programme of strong Artificial Intelligence.

As argued in this chapter, it is a requirement which has foundations in the age-old philosophical and civilizational development. Thus Leibniz's name appears as a symbolic bridge between this splendid tradition and the nascent civilization in which automated information-processing was to become a decisive factor for further development.[10]

3.3.4. When concluding this chapter and hinting at its relations to both preceding and following ones, I take advantage of the phrase 'symbolic bridge' as referring to Leibniz above. Leibniz's ideas deserve to be extensively commented. However, the problem of their historical influence is a matter that must be handled in a sophisticated manner. Only his programme for mathematical logic and his idea of the universal formalized language were known to posteriority, while his results themselves were not discovered until the end of the 19th century.

Yet the very fact that Leibniz's programme was being realized without Leibniz, up to the present stage of Artificial Intelligence, is instructive and thought-provoking. The intellectual forces acting on behalf of this development proved so strong that even the loss of Leibniz's so significant contribution, caused by the delay over two-centuries of its being published, did not hamper the progress. Leibniz was the herald of the coming epoch, but one whose voice had not been heard until the time in which his visions started to

[10] In this chapter attention is paid to Leibniz's contributions to mechanization of deduction, but his role of a bridge should be seen in other domains of logic as well, esp. modal logic. As to the latter, there are many thorough studies to render Leibniz's modal logic in terms of certain present-day systems, e.g. (Lenzen 1987), which offers numerous references.

be materialized, and his results anew achieved by others — with limitations whose unavoidability has in the meantime been recognized.

This is a precarious position, indeed. The assessment of Leibniz's position therefore requires a sophistication which does not attribute too much to him, and at the same time is able to acknowledge the great role he nevertheless played — as a symbol of epoch-making ideas and of the civilizational turn.

There is a wonderful counterpoint between the Leibnizian and the Cartesian approaches to mind and logic. Even Leibniz's drawbacks contribute to our understanding of the mind when confronted with the views and approaches of Descartes and Pascal. When his belief in the power of mechanized thought proves unrealistic, we are able to answer the question 'why?' with the help of other protagonists. Descartes' and Pascal's preoccupation with the mind's inner life counterbalances Leibniz's tendency toward replacing mental acts by operations on symbols which could be formalized and mechanized. In the Cartesian-Pascalian perspective we see the mind as dealing with objects and relations which are so numerous, subtle and involved that in many cases a verbalization or symbolization must prove inadequate.

In this chapter, certain programmes and ideas which do not prove to have had any bearing on the mechanization of reasoning have been also dealt with at some length. What Ramus, Melanchton, and even Bacon, designed and postulated for logic in their time died away without any impact on the further progress of logic. Nevertheless, the story of their endeavours and dreams undoubtedly deserves to be told, for we can learn from it which intellectual forces are fit enough to shape human thought and civilization. And against such a background, the main protagonists of the process to be studied can be seen and appreciated all the better.

CHAPTER FOUR

BETWEEN LEIBNIZ AND BOOLE: TOWARDS THE ALGEBRAIZATION OF LOGIC

4.1. Preliminary remarks

4.1.1. The algebraization of logic initiated by Leibniz marked the decisive step towards the mechanization of arguments. The equally important step which preceded it consisted in the introduction of variables combined with the conception of logical consequence as dependent solely on the logical form as, in turn, dependent solely on logical constants. But that was not so much a single step as the age-long process from Aristotle to the late scholasticism. The next turning point after algebraization is to be seen in the invention by Frege and later Russell of the calculus of quantifiers, which turned logic into a universal instrument of reasoning, yet at the price of what – if we follow Hilbert (1925) who defined the axioms of that calculus as transfinite – might be termed the infinitism of logic. In order to return to algebraic equations which offered the possibility of mechanical combinatorics performable by the computer it was necessary to find methods of eliminating the quantifiers, that is to do something which might in turn be termed the finitization of first order logic. As is known, that was achieved owing to Skolem, Gentzen, Herbrand and Beth.

Thus the mechanization of logic proceeded in the three-stages development consisting of formalization (in the sense of basing logic on the concept of logical form), algebraization, and infinitization (intended to increase the strength of proofs) combined with the possibility of reducing the latter to a finitized form that can be processed mechanically.

4.1.2. The epoch of algebraization spread from Leibniz to Boole and his successors. The present chapter is concerned with that epoch treated as an open interval, that is without Leibniz (discussed in the preceding chapter) and without Boole (to be discussed in the next one). That temporal interval, however, does not form a genetic sequence which would be formed by the relation 'x influenced y'. The programmes of treating logic as an algebraic calculus and the implementations of those programmes (only fragmentary before Boole) turn out to have been largely independent of one another. In particular, there are no traces of Leibniz's influence upon the 18th century authors, because, as is now commonly known, he did not publish his results, which he regarded as incomplete; thus his notes remained buried in the archives of the library in Hannover until the moment of being unearthed by Couturat.

Thus the historian of that period is not very likely to enjoy discovering in it how some authors influenced other ones. But he may compensate that by enjoying a confirmation of what can be called the law of the *objective development of ideas.* That law manifests itself in the well-known phenomenon of the simultaneous or almost simultaneous independent appearance of similar discoveries or inventions. That phenomenon can be interpreted as a result of the evolution of the collective consciousness of a certain milieu (consisting of researchers, their readers and critics, sponsors, etc.), consciousness which is, as it were, similarly programmed by the problems posed in a given period, the state of knowledge, the state of language (conceptual apparatus), and research methods.

It is precisely that phenomenon of the objective development of ideas that we have an opportunity to watch when studying the development of logic between Leibniz and Boole. That stage, still little known at the time of the pioneer studies in the history of logic initiated in the 1930's, became the subject matter of intensive investigations in the past thirty years (largely owing to the excellent series of reprographic editions of old masters). The way to such research has been paved by thorough factfinding studies of such authors as Hans Werner Arndt and Wilhelm Risse. Now that so much textual work has been done, the task of next researchers is to pose new question addressed to those texts.

The present chapter consists just in posing such a new question provoked by the development of computational logic treated either as an instrument for the development of computer science or as developing itself with the help of computers. It is the latter point of view which in adopted in the present historical study.

4.2. Leibniz's direct successors

4.2.1. The Swiss mathematicians Johann Bernoulli (1667–1748) and his brother Jakob Bernoulli (1654–1705) are sometimes called Leibniz's disciples, which, however, is not to be understood in the sense of different generations (Leibniz lived in 1646–1716) nor in the sense of their having been his students or members of a school of him. The most important form in which Leibniz influenced others was rather his personal correspondence than publications, and than lectures (which he did not hold as he gave up a standard academic career). In this case, too, we have to do with influence through correspondence, and hence due to a certain relation of partnership rather than with the asymmetric relation between a teacher and his disciples. But it is worth noting that the latter relation did hold between Johan Bernoulli and Leonard Euler (1707–1783), one of the main heroes in the history of the algebraization of logic, who not only attended Bernoulli's lectures at the university in Basle but also availed himself of that kind of personal acquaintance with him which was a typical relation between disciple and master.

The brothers Bernoulli were also interested in logic, which fact resulted in the book *Parallelismus ratiocinii logici et algebraici* (1685). Here are the statements to grasp that parallelism (quoted after Risse 1970, p. 258 f., ad hoc translated from Latin by W. M.).

> When ideas of many things are combined, that is done by means of the copula; and when ideas of many possibilities are combined, that is done by means of the sign +. When two ideas are affirmed or denied about one another through the use of particles *is* or *is not*, that is called a statement (*enuntiatio*). When two quantities [...] are linked together by the sign of equality =, that is called equation [...] whereas inequality is denoted by the signs < and >.

Such observations still do not go beyond Leibniz's ideas who, e.g., in his *Generales inquisitiones de analysi notionum et veritatum* vastly expanded logic as an algebraic calculus, and at the same time believed that the logic inherited from Aristotle and the schoolmen was an unavoidable cognitive instrument that considerably gains from its combining with an algebraic calculus but has a significant value in itself. It is otherwise in Bernoulli's case. When considering his other statements, we find that he appreciated algebra more than logic. The former exceeds the latter by its generality and the range of applications, and hence can take over the most important task programmatically set to logic in the 17th and the 18th century, namely the role of an instrument in the discovery of the truth. Here is the text which voices that conviction.

> [...] the analytic art otherwise called algebra. This is that great art of discovery which leads the mind by the admirable method, with the utmost mastery, and by reliable and infallible procedure, to the cognition of what is sought. It is as much superior to ordinary arithmetic as the latter exceeds popular logic.

Bernoulli termed algebra divine mathematics, which succeeds where ordinary logic proves helpless. Here is his Baroque apostrophe:

> Hence come divine mathematics [...] thou beginst where popular logic ceases. Thou art more perspicacious than Argo where the other proves blind.

4.2.2. Christian Wolff (1679–1754) was an unquestionable disciple of Leibniz. He assimilated the views of his master following the study of his works and intensive correspondence. Note that it was under Leibniz's influence that the young Wolff abandoned the Cartesian view of logic, which was marked by a negative assessment of all formalism, especially the syllogistic one, and postulated the pursuit of logic as a normative theory of controlling the mind, with the disregard of the role of language. From that time on Wolff became an ardent adherent of syllogistic and the programme of pursuing it in the algebraic way as operations on symbols alone, whose meanings might be entirely disregarded. Wolff himself did not contribute to the implementation of that programme of algebraization of logic, but in view of the scope and strength of his influence one may assume that his agitation helped to consolidate that state of consciousness which ultimately brought the perfect rendering of logic in algebraic terms.

Johann Lambert, who rendered the greatest services to the algebraization of logic in the 18th century, worked under Wolff's influence, but before we discuss his ideas, it is in order to mention at least two of his contemporaries, with whom he engaged in lively discussions. That review, even if we disregard other representatives of the trend furthering the algebraization of logic (and they were quite numerous in that century), can help one to form an opinion on the vigour and scope of that trend.

4.2.3. Gottfried Ploucquet (1716–1790), a German author of French origin, was busy in developing the Leibnizian ideas of universal language (*characteristica universalis*) and logical calculus (*calculus ratiocinator*). He studied diligently the then accessible logical texts of Leibniz, including *Difficultates quaedam logicae*. Ploucquet as the author of many works written in German and engaged in public discussions with Lambert, was well known to 19th century German philosophers. He was quoted, although with disapproval, by Hegel, which contributed to his being known in the next century. Hegel was twenty years old when Ploucquet died, so they were almost contemporaries; this may be interpreted as a symbol of the continuity of logical problems in German philosophy from the 17th to the 19th century. Thus while Leibniz's logical calculi themselves remained unknown, his seminal idea of logical calculus as a certain algebra inspired later authors, partly because of Leibniz's authority, and partly because it was anyway 'floating in the air' due to the atmosphere of the admiration for mathematics, to the impression made by the successes of algebra, and to the still strong position of the logical tradition inherited from the Middle Ages. That idea was expressed by the title of one of Ploucquet's numerous works, namely *Methodus calculandi in logicis* (1763).

Ploucquet can well be used as a witness in the trial of Ramon Lullus (see Chapter 2 of this book) or rather those who spread the legend that he had been a forerunner of the mechanization of arguments. In fact, there is in Lull no attempt to prove logical theorems, and *a fortiori* there are no algorithms whereby such a proof could be constructed. On the other hand, from Albert the Great to Tartaret we encounter the search for such algorithms for syllogistic (as to Tartaret's *pons asinorum* see Chapter 2, Sec. 2.6, above). The fact that syllogistic constitutes a small fragment of logic and that it was erroneously identified with logic as a whole was not an obstacle to the effectiveness of such a search. That fragment was well suited as the experimental field for the algorithmization of proofs, and further the fact that the medieval endeavours were naive and little successful was due to the inability of the early authors to combine them with the calculatory approach. But two things were essential: the formulation of the problem and the proper establishing of the area of research. When in the 17th century algebra provided the paradigm that could be used in logic, a new and promising field of research was opened. It is in that field that we find Ploucquet as its typical representative.

Ploucquet tested the efficiency of his algebraic symbolism on the language of syllogistic. For the recording of statements forming the so-called square of opposition he adopted a new syntax, in which he succeeded symbolically to express both a kind of the quantification of both the subject and the predicate, which made it possible to render the structure of a categorical sentence by juxtaposition alone — as in algebra, where juxtaposition happens to be a symbol of some operation. Thus, by drawing from the traditional language the symbols S and P to denote, respectively, the subject and the predicate he makes a distinction between any of those terms being taken in its whole extension and its being taken in a part of its extension, the distinction being indicated, respectively, by a capital and a lower-case latter. The distinction between distributed and not distributed terms was a scholastic invention on the path to the algorithmization of syllogistic (admired by Leibniz), and hence we have here to do with the scholastic idea in a new symbolic attire. Here is a text on that matter chosen by way of example (Ploucquet 1766, p. 63 f.; quoted after Riesse 1970, p. 283, ad hoc translated from Latin by W. M.).

> The method of calculating syllogisms consists in that the proposition by which the middle term is taken universally is placed first, and the other is connected to is so that the middle term is placed in the middle. Thereby the middle term is deleted and the inferred conclusion is necessarily demonstrated. And when the middle term is universally taken twice, the order of its position is indifferent. Let us consider, for instance, Mp, Sm. When m is subsumed under M, if pM is true then pm is true, too, and hence pmS is written; when m is deleted, we have pS or Sp, which is to say that all S is some P, or all S is P.

This text, deliberately setting up a syllogistic calculus, should be interpreted in the light of the explanation given several pages earlier (Ploucquet 1766, p. 57) and pertaining to the identity of the extension of subjects and predicates in affirmative sentences. Such an interpretation of the structure of the categorical affirmative sentence proves inseparable from the quantification of predicate.

> Since in every affirmation the identity of the subject with the predicate is asserted, it is obvious that if the subject is denoted by S and the predicate by p, then p can be substituted for S, and reciprocally, S for p. Thus Sp is identical with pS.

It can be seen from this text that should we, as it is customary today, interpret Sp in the quantification calculus as the implication

$$\forall x[S(x) \Longrightarrow p(x)],$$

and not as the equivalence

$$\forall x[S(x) \Longleftrightarrow p(x)],$$

that would be at variance with the reversibility (reciprocality) stated in the text quoted above. This is a warning against the interpretation of traditional logic in terms of the quantification calculus with disregard of the context of the old traditional theory. This is so because in those theories which use the algorithm of the recognition of the correct conclusion, based on the concept of distribution of terms, the predicate in a general affirmative sentence has as its extension a certain subset of the set denoted by that predicate in other contexts (or outside of context) — precisely that subset which coincides with the extension of the subject. Now we can follow the reasoning in the text quoted earlier having in view the fact that with our author juxtaposition is a syntactical indicator of affirmative sentences, whereas for negative ones he used his own symbol of the relation of exclusion, namely $>$.

This is reasoning for testing the correctness of syllogism, which is an algorithm that refers solely to syntactic features, that is the shape and position of symbols, as arithmetical algorithms do. We have to do with the following rules. (1) It is allowed to transpose letters. (2) It is allowed to replace a capital letter by a lowercase one. (3) Given a sequence of pairs of letters such that the consequent in every pair is identical with the antecedent of the next one it is allowed to accept the sequence of letters consisting of all antecedents and the last consequent, for instance to pass from ab, bc to abc (to be read: a is b and c, for instance from "the lion is a cat" and "the cat is predatory" we pass to "the lion is a predatory cat". (4) In such a sequence of letters (as that described above) it is allowed to omit any letter except for the first and the last one.

After such a reconstruction of Ploucquet's procedure it can better be understood why he considered it as a procedure analogical to the calculus carried out on symbols. Its discouraging intricacy largely explains why, even though Ploucquet's works were popular in his times, he did not become the Boole of the 18th century. He lacked bolder ideas of his own, and instead of that he seems to have pedantically changed into small pieces what reached him from Leibniz's ideas. Nevertheless he is a significant witness of his epoch, in which syllogistic played an important role as the experimental field for the endeavours to transform logic into an algebraic calculus.

The same trend embraced Georg Jonathan Holland, an adherent of Ploucquet in the latter's polemic with Lambert. Without going into the details of Holland's views and that polemic which was much publicized in that time, it is worth while quoting the title of his fundamental work which testified to the then typical linking together of mathematics, extra-mathematical calculi, and the programme of a universal symbolic language. The title reads *Abhandlung über die Mathematik, die allgemeine Zeichenkunst und die Verschiedenheit der Rechnungsarten* (1764).

4.3. The work of J. H. Lambert

4.3.1 We shall discuss here the merits of Johann Heinrich Lambert (1728–1777) for the algebraization of logic. Even though that work is imposing by its size, also in the sense of the number of comprehensive volumes, it is a small and even as it were marginal element in the vast scientific production of that son of a tailor in Mülhausen in Switzerland. Lambert was a self-taught person who first earned his living as a clerk and later as a private teacher before his two treatises on optics (1759) brought him publicity and posts in the Bavarian Academy of Sciences and later, until his death, in the Prussian Academy of Sciences.

Lambert's works on logic fill at least seven volumes (accessible in reprographic editions). From the present point of view the most important ones are *Neues Organon oder Gedanken über die Erforschung und Bezeichnung des Wahren* (1764) and *Logische und philosophische Abhandlungen* (1782). The latter work is a collection of papers published posthumously by the Prussian Academy in Berlin and Johann Bernoulli personally as the editor of Lambert's writings. The editorial introduction to the work, written by Christoph Heinrich Müller, Lambert's compatriot and friend, deserves a separate mention because of a certain conception of the mechanization of arguments. That conception must have been sufficiently interesting and comprehensible to the learned readers of that period, which is borne out by the fact that such an experienced editor as Müller (who was busy publishing German literature from the period 1050–1500) commented on it in such a way as the problem were something being a vogue.

When we trace references to the literature of the subject in Lambert's works on logic we find that few of them refer directly to Leibniz. Lambert was directly influenced above all by Locke and Wolff. From the works of the latter Lambert learned about Leibniz's programme of the mathematization of logic, but not about any definite calculus, because Leibniz's endeavours in that field were not known to Wolff either. That lack of information in the case of Wolff is hardly comprehensible because his correspondence with Leibniz continued for many years and gave him many opportunities to ask the master about his logical ideas. Note that Wolff used to announce that it was to Leibniz that he owed both his belief in the value of syllogistic (which he had earlier treated disparagingly under the influence of Cartesianism) and in the programme of its mathematization. If we consider that Wolff was a competent mathematician, valued and recommended for professorship in Halle by Leibniz himself, we find it even more difficult to understand why we do not encounter in Wolff's vertiginously great production any endeavours to algebraize logic. This provides us with a convincing proof of Lambert's independence and originality in his calculatory implementations of Leibniz's programme preached (but not practised) by Wolff.

That fact is not obvious as yet to everyone as even competent authors claim that Lambert merely developed (and not rediscovered himself) Leibniz's results. For instance, Styazkin (1969), when writing about linear diagrams, whose authorship Lambert ascribed to himself *expressis verbis* (he could not have known Leibniz's *Generales inquisitiones*), formulated such an opinion: "Evidently following Leibniz's method, Lambert proposed a geometric interpretation of syllogistic functors." (p. 121).

4.3.2 Before we proceed to present relevant solutions suggested by Lambert in greater detail it is to the point — in order to obtain a proper perspective — to revert to the historical relationship between algebra, as developed in the 16th and the 17th century, and traditional syllogistic. This issue recurs in variations like a refrain in this book because it is the key to decipher the process whereby logic was developing towards the mechanization of arguments.

We are here to do with a case in which the intersection, at a certain point of time, of two mutually independent processes opens a new stage in history. Such intersection may also appear in the fact that a certain group of scholars is engaged with similar intensity in both processes. This was just in the case under consideration: people strongly rooted in the tradition of scholastic logic came to be interested in algebra, and some of them proved creative in that newly born discipline called then *logistica speciosa universalis*, that is (in a free translation), the general theory of calculating with variables (i.e., sign denoting *species* of objects, e.g. numbers, instead of individual objects). The prominent personalities active on that point of intersection self-evidently included Leibniz and Lambert, hence we find it easy to understand some similarities between them in spite of the fact that Lambert did not know Leibniz's results.

Two ideas were drawn from the logic of Aristotle and the schoolmen. Each of them entered into a natural relationship with a certain algebraic concept in such a way that all components came to form a coherent whole. One of those logical ideas was already discussed more extensively in Chapter 2 (Sec. 2.5). It is the problem, posed already by Aristotle, of finding the middle term as a means of constructing the proof. The problem consists in the fact how that still unknown middle term is to be found on the basis of what is known, namely the external terms and the intended conclusion.

This task was generalized as the problem of finding new data of a definite kind on the basis of given data of a definite kind. The logic conceived as the theory of such a search was termed *logica inventionis* which can be rendered as inventive logic, or (more freely) logic of discovery.

Thus generalized programme of inventive logic was characteristic of the Renaissance rather than of the Middle Ages, and had its standard-bearer in Francis Bacon (for more details see Chapter 3), but the point of departure should be seen in the Middle Ages; a story related to that problem can be traced in Leibniz's dissertation *De arte combinatoria.* Moreover, many scholars who combined interests in logic and algebra must have then been impressed by the analogy between such a finding of new data and the finding of the unknown in an algebraic equation.

But if use was to be made of that analogy, the sentences considered in logic had to have a form analogical to that of equations. That would also result in the attractive possibility of constructing proofs by replacing one side of an equation by an expression equivalent to it (this was how the nature of the proof was conceived by Leibniz). But it is well known that none of the four types of categorical sentences of Aristotelian logic possesses the form of an equation. We shall see how the inspirations derived from medieval logic helped to solve that difficulty.

The theory of suppositions made medieval logicians used to thinking that what determines the logical function of an expression is not only its external form and the related lexical meanings, but also its context, and the intention of the speaker marked in some way. Hence, according to the context in question, an expression one and the same in its form could stand for an individual, either definite or indefinite; Latin did not make any formal distinction in that respect as it lacked articles; in another context a name could refer to a class of individuals, to a universal, etc. Upon that habit of thinking there was imposed the theory of the distribution of terms in categorical sentences. Taking a general term on one occasion in its full extension and on another occasion only in a part of that extension would amount to different suppositions. The latter of the two, that is the use of a term as denoting a subset (of the set which is the extension of a given term in its lexical meaning) can be indicated by a separate word (for instance *aliquis*, i.e., 'some') or by the position in the context of a definite sentential structure, for instance, the position of the predicate in a general affirmative sentence (customarily denoted by SaP).

There was the last step to be made. We shall discuss it by taking SaP as an example. In contemporary logic, when traditional logic is interpreted in terms of the predicate calculus, SaP is treated as a record of an inclusion because both terms in the sentence are taken in their full extensions. But if P is taken in a part of its extension, namely that which coincides with the extension of the subject, than SaP is expressing an equation in which *est* should be interpreted as '='. In this way we arrive at a form comparable with the form of an algebraic equation. It is worth noting that for the logicians who lived in the 17th and the 18th century that interpretation was so obvious that they did not think it necessary to explain the matter to the readers, which they would have to do if they addressed their texts to the 20th century readers (who have forgotten the old teachings on suppositions and the distribution of the syllogictic terms).

4.3.3 The information on the algebraic aspect of Lambert's logic should accordingly begin with his declaration on the role of the relation of identity between concepts. We find it, among other things, in his work *Anlage zur Architectonic* (1771, p. 90; quotations after Risse 1970, p. 271, ad hoc translated from German by W. M.).

> The concept of identity is, from the point of view of the fundamental theory, what the concept of equality is from the point of view of mathematics. We cannot draw any conclusion without being certain of the identity of the middle terms in both premisses and of the identity of the external terms in the premisses and the conclusion.

The following formulation, found in the previously mentioned *Logische Abhandlungen* (p. 79; see Risse 1970, p. 269), tells us about the relation between identity and the main task of logic conceived as *logica inventionis*.

> Logical analysis is an art of deducing unknown or sought concepts from known or given ones by means of identities. Since nothing can be found out of nothing, concepts have to be

> given to allow one to find the unknowns. ... General analysis *(analytica logica speciosa, logistica speciosa universalis)* is an art of deducing concepts from general and indefinite ones. Since the concepts needed for that purpose are indefinite, one cannot use words as these denote definite concepts. The most appropriate thing is, therefore, to use letters and other signs as it is also the case in algebra.

Here is an example of a certain technical solution which makes it possible to impart to a sentence of the type SaP the form of identity:

$$S = xP,$$

where juxtaposition is the logical equivalent of the algebraic operation of multiplication, and the symbol x stands for an object which, when multiplied by P, yields an object identical with S. For instance, the sentence, "Every Greek is intelligent" will be interpreted as the identity:

> The set of Greeks is identical with a certain subset of intelligent beings.

That subset is obtained by the intersection of the set of intelligent beings with some other set, in the above formula represented by x, that is, according to Lambert's expression, an indefinite coefficient. Thus, the syllogism Barbara is reconstructed as the following set of identities:

$$M = xP, \qquad S = yM.$$

On replacing M in the second premiss by xP we arrive at the conclusion:

$$S = yxP.$$

These few specimens of algebraic treatment of logic in the 18th century should provide us with a background at which we can clearer perceive the developments in the next centuries — those forming the proper history of mechanization of reasoning. Before we discuss them in the chapters which follow, let the story of incubation period, that ranging from the Middle Ages to Lambert's projects, be closed with the following moral.

4.3.4. As commonly known, the first mature systems of algebraically moulded logic were due to British mathematicians and philosophers, while the preceding development occurred at the Continent. Moreover, there is the case, too rarely contemplated, that those British authors moved against the steady stream of the empiricist philosophy and logic, so dominating on the British intellectual scene since Francis Bacon. Thus we come to the following question: whether those algebraically-minded logicians were dissidents influenced by foreign ideas, or did they hit upon their ideas by themselves? The question is vital, as the answer should throw light on some fundamental facts concerning the development of ideas.

That answer is expected from a fact-finding historical research. Such a research was done by Peckhaus (1994) who found out important details to shed light on the issue of the British-Continental relations in developing logic. His conclusion is to the effect that Boole and his British collegaues first discovered the algebra of logic, and then learned that it has been already discovered by Leibniz. This is not to mean that the same was made twice; in Leibniz one finds just some primordial ideas. Anyway, the consciousness of anticipating those achievements by Leibniz proved a significant circumstance in the further development.[1]

The statement of that significance may seem strange, since the lack of any causal nexus between a forerunner and a genuine originator, in the case of Leibniz and Boole confirmed by most recent investigations (cf. Peckhaus 1994), should imply that the history would have developed in the identical way even if the forerunner had never existed. However, this is not the case in the history in question, which can be explained by a kind of feedback between own results and their later detected anticipation. Boole and the other British logicians, including Jevons (specially important because of the step he made from algebraization to mechanization) not only produced technical results but also entertained some philosophical ideas which did not conform to the empiristic trend, characteristic

[1]The problem of how much of modern logic, esp. algebra of classes, was anticipated by Leibniz belongs to main issues in Leibniz scholarschip. In this book a rather conservative attitude is assumed, for bolder interpretations require more research. Among those who pioneer such a bolder approach there is Wolfgang Lenzen. See, e.g., (Lenzen 1984), and other works (as reported in Leibniz Bibliography in *Studia Leibniziana*).

of the British philosophy. It required a strong belief that the laws of thought were independent from empirical reality, in order to overcome the pression of the British empiristic orthodoxy; that belief gained a new vigour, due to its meeting with Leibniz's tenets, in the period in which it must have fought for recognition and standing in the British philosophy (had not such a recognition become the case, there would not have been the way paved to Russell's (1900) seminal work on Leibniz).

For instance, William Stanley Jevons (1874), contrary to the teachings of John Stuart Mill, William Whewell, etc., tried to deductively establish the scientific method on the basis of Boole's logical principles (again, cf. Peckhaus 1994, p. 592). As for Boole himself, he voiced his conviction that the laws of logic have a real existence as laws of the human mind, independent from any observable facts (contrary to the creed of the British empiricism). As commented by Peckhaus (1994, p. 590), such statements of Boole (1854, p. 40) prove his affinity with Leibniz. This may be taken as a proof of reality of what in German is called *Geistesgeschichte*, that is the history of objective ideas which must come into being, due to their internal logic, in a way which is relatively independent from biographical circumstances.

Owing to that independence, there are posssible purely intellectual meetings across centuries as that between Boole and Leibniz, described by Boole's widow Mary Everest Boole.[2]

> Some one wrote to my husband to say that, in reading an old treatise by Leibniz (who lived at the same time as Newton) he had come upon the same formula which the Cambridge people call "Boole's equation." My husband looked up Leibniz and found his equation there, and was perfectly delighted; he felt as if Leibniz had come and shaken hands with him across the centuries.

[2]See Boole, M. E. 1905, p. 1142; quoted after (Peckhaus 1994, p. 591). The equation mentioned amounts to the formula $aa = a$, called by Boole law of duality, which was highly appreciated as non-trivial (because of its being at variance with arithmeticians' habits) and as being, according to Boole's (1847) own words, the distinguishing feature of his calculus (cf. Kneale and Kneale 1962, p. 408).

CHAPTER FIVE

THE ENGLISH ALGEBRA OF LOGIC IN THE 19TH CENTURY

The aim of this chapter is to study works and achievements of English logicians of the 19th century. We shall consider the role they played in the development of mathematical logic, in particular their contribution to the formalization of logic and to the mechanization of reasonings. We shall present and discuss first of all works of A. De Morgan, G. Boole, W. S. Jevons and J. Venn indicating their meaning and significance for the development of mathematical logic.

Works of English logicians of the 19th century grew out of earlier ideas and attempts of G. W. Leibniz, G. Ploucquet, J. H. Lambert, L. Euler (cf. previous chapters). The idea of the mathematization of logic and the development of the formal algebra in the 19th century were sources of the algebra of logic established by De Morgan, Boole and Jevons. It was in fact the beginning of the mathematical logic. The old idea of a logical calculus which would enable the analysis of logical reasonings with the help of a procedure similar to the procedure of solving equations in algebra was realized.

5.1. De Morgan's syllogistic and the theory of relations

5.1.1. We shall begin the discussion of the development of logic in England in 19th century by studying the idea of quantification of the predicate.

In traditional logic (since Aristotle) the most important role was played by syllogisms. Aristotle defined them as formal arguments in which the conclusion follows necessarily from the premises. His analysis centered on a very specific type of argument. He considered namely statements of the form: all S is P (universal affirmative), no S is P (universal negative), some S is P (particular affirmative), and some S is not P (particular negative), abbreviated later, resp., as follows: SaP, SiP, SeP, SoP and called the categorical sentences. Aristotle observed that one can build valid schemas of inference consisting of two premises and a conclusion being categorical sentences – they are called categorical syllogisms. If we assume that every term in a syllogism stands for a nonempty class then we get that 24 of 256 possible combinations are valid inferences.

5.1.2. Aristotle and his medieval followers greatly exaggerated the importance of the syllogism. Nevertheless syllogism formed the main part of logic until the beginning of mathematical logic in 19th century. Various attempts to reshape and to enlarge the Aristotelian syllogism were undertaken. Let us mention here attempts of F. Bacon, Ch. von Sigwart and W. Schuppe. The most famous of those attempts was the "quantification of the predicate" by the Scottish philosopher Sir William Hamilton (1788–1856) presented in his book *Lectures on Metaphysics and Logic* published in 1860. He noticed that the predicate term in each of Aristotle's four basic assertions SaP, SiP, SeP, SoP is ambigous in the sense that it does not tell us whether we are concerned with all or part of the predicate. Hence one should increase the precision of those four statements by quantifying their predicates. In this way we get eight assertions instead of Aristotle's four, namely:

all S is all P,
all S is some P,
no S is all P,
no S is some P,
some S is all P,
some S is some P,
some S is not all P,
some S is not some P.

Using those eight basic propositions we can combine them to form 512 possible moods of which 108 prove to be valid. The usage of statements with quantified predicates allows us higher precision than it was possible before. For example, the old logic would treat "All men are mortal" and "All men are featherless bipeds" as identical in form; whereas in the new system we see at once that the first statement is an example of "All S is some P" and the second is an example of "All S is all P". But there arose some problems. It was difficult to express those new statements with quantified predicates in a common speech. Without developing a really complete and precise system of notation one finds himself forced to emply words in a clumsy and barbarous way. Hamilton was aware of it and attempted to remedy the obscurity of phrasing by devising a curious system of notation. Though it was really curious and rather useless in practice it was important for two reasons: it had the superficial appearence of a diagram and it led Hamilton to the idea that by transforming the phrasing of any valid syllogism with quantified predicates it may be expressed in statements of equality. The latter suggested that logical statements might be reduced to something analogous to algebraic equations and so gave encouragement to those who were seeking a suitable algebraic notation. Some logicians are of the opinion that this was Hamilton's only significant contribution to logic (cf. Gardner 1958).

5.1.3. The idea of quantification of the predicate in the syllogism can be found also in papers of another English logician Augustus De Morgan (1806–1871). He was a mathematician (contrary to W. Hamilton) and worked not only in logic but also in algebra and analysis. We shall not discuss his mathematical works here but concentrate on his logical achievements. Nevetherless we want to mention that it was just De Morgan who introduced the notion "mathematical induction" which was popularized later thanks to the book on algebra written by I. Todhunter (1820–1884).

A. De Morgan's earlier logical works (i.e. works written before 1859) were devoted to the study of the syllogism. He introduced independently a system more elaborated than Hamilton's one. Hamilton accused him of plagiarism and for many years the two men argued with each other in books and magazine articles. It was, as M. Gardner (1958) writes "perhaps the bitterest and funniest debate about formal logic since the time of the schoolmen, though most of the humor as well as insight was on the side of De Morgan". This debate had also serious consequences. Namely it caused G. Boole's renewed interest in logic which led him to write the book *Mathematical Analysis of Logic* published in 1847 (we shall discuss it in the next section).

Trying to reshape and to enlarge the Aristotelian syllogism (cf. the book *Formal Logic, or the Calculus of Inference Necessary and Probable* 1847) De Morgan observed that in almost all languages there are so called positive and negative terms. Even if in a language there are no special words indicating this dychotomy, nevetherless every notion divides the universe of disourse into two parts: elements having properties indicated by the given term and those which do not have those properties. Hence if X denotes a certain class of objects then all elements of the universe which are not X can be described as not-X. The latter is denoted by De Morgan by x. In this way the difference between positive and negative terms disappears and they are posessing equal rights. This enabled De Morgan to consider – instead of two terms of the traditional syllogistic X, Y – four pairs of terms: X, Y; x, y; X, y; x, Y which give 16 logical combinations, 8 of which are different. He introduced various types of notation for them. The first two consisted of letters and resembled the traditional notation: a, i, e, o; the latter two consisted of systems of parentheses. We shall present them in the following table (the notation in the last column was introduced in De Morgan 1856, p.91):

all X is Y	A	A_1	$X)Y$	$X))Y$
no X is Y	E	E_1	$X.Y$ or $X)y$	$X).(Y$ or $X))y$
some X is Y	I	I_1	XY	$X()Y$
some X is not Y	O	O_1	$X:Y$ or Xy	$X(.(Y$ or $X()y$
all x is y	a	A'	$x)y$ or $Y)X$	$x))y$ or $X((Y$
no x is y	e	E'	$x)Y$ or $x.y$	$x))Y$ or $x(.)Y$
some x is y	i	I'	xy	$x()y$ or $X)(Y$
some x is not y	o	O'	xY or $x:y$ or $Y:X$	$x()Y$ or $x).)Y$

We see that De Morgan's symbols introduced to express the old and new types of syllogism were not algebraic. But he found that his symbols can be manipulated in a way that resembled the familiar method of manipulation of algebraic formulae (cf. De Morgan's letter to Boole from 16 October 1861 – Letter 70 in Smith 1982, pp.87-91).

The above statements were called by De Morgan simple. He established the following connections between them:

$$\begin{aligned}
&X)Y = X.y = y)x\,;\\
&X.Y = X)y = Y)x\,;\\
&XY = X:y = Y:x\,;\\
&X:Y = Xy = y:x\,;\\
&x)y = x.Y = Y)Y\,;\\
&x.y = x)Y = y)X\,;\\
&xy = x:Y = y:X\,;\\
&x:y = xY = Y:x\,.
\end{aligned}$$

The sign = was used by De Morgan in two meanings. Here it means "equivalent" but in other contexts it was used also as a symbol for "implies".

Denoting by + the conjunction De Morgan established the following complex statements:

$$\begin{aligned}
&D = A_1 + A' = X)Y + x)y\,;\\
&D_1 = A_1 + O' = X)Y + x:y\,;\\
&D = A' + O_1 = xy + X:Y\,;\\
&P = I_1 + I' + O_1 + O'\,;\\
&C = E_1 + E' = X.Y + x.y\,;\\
&C_1 = E_1 + I' = X.Y + xy\,;\\
&C' = E' + I_1 = x.y + XY.
\end{aligned}$$

Using the contemporary set-theoretical symbolism we see that D is simply $X = Y$, D_1 is $X \subset Y$, D' is $X \supset Y$, C is $X = y$, C_1 is $X \subset y$, C' is $X \supset y$, P means that neither $X \cap Y$, nor $X \cap y$, nor $x \cap Y$ nor $x \cap y$ is empty.

Having introduced those fundamental relations De Morgan developed his theory of syllogism. He considered not only syllogism with simple premises but also with complex ones (which he called complex syllogism).

We shall not present here all details of De Morgan's syllogistic. Let us note only that he used simple terms X, Y etc. as syllogistic terms as well as complex ones: PQ (being P and Q; $P \cap Q$ in set-theoretical terms), $P * Q$ (being P or Q or both; $P \cup Q$), U (the universe), u (empty set). He stated the following properties:

$$XU = X\,; \quad Xu = u\,; \quad X * U = U\,; \quad X * u = X\,.$$

Having introduced operations which we call today the negation of a product and sum he established the following connections (called today De Morgan's laws):

$$\text{negation of } PQ \text{ is } p * q,$$
$$\text{negation of } P * Q \text{ is } pq.$$

He noted also the law of distributivity:

$$(P * Q)(R * S) = PR * PS * QR * QS\,.$$

We see that developing the theory of syllogism De Morgan came to the idea of what we call today Boolean algebra. He did it independently of G. Boole (in the next section we shall discuss in details connections and interdependences of their works and compare their achievements).

De Morgan not only extended the traditional syllogistic but introduced also new types of syllogisms. In *Budget of Paradoxes* (1872) he summarized his work under six heads, each propounding a new type of syllogism: relative, undecided, exemplar, numerical, onzymatic and transposed (cf. also De Morgan 1860). Since those considerations did not lead to new developements in mathematical logic we shall not go into details here.

5.1.4. De Morgan's considerations of syllogism were not very original. Much more original were his later works – and it is perhaps his most lasting contribution to logic. In the paper "On the syllogism no IV, and on logic in general" (De Morgan 1861) from 1859 (published in 1864) he moved beyond the syllogism to investigate the theory of relations. Although he was not the first to study relations in logic, he was probably the first who gave this subject a concentrated attention. He may be considered as a real founder of the modern logic of relations. His ideas and conceptions were later on developed by various logicians, first of all by Ch. S. Peirce, E. Schröder, G. Peano, G. Cantor, G. Frege and B. Russell. Ch. S. Peirce wrote that De Morgan "was one of the best logicians of all time and undoubtedly a father of the logic of relatives" (cf. Ch. S. Peirce 1933-34, vol.3, p.237).

In "On the syllogism no IV" De Morgan noted that the doctrine of syllogism, which he had discussed in his *Formal Logic* of 1847 (cf. De Morgan 1847) and in earlier papers was only a special case in the theory of the composition of relations. Hence he went to a more general treatment of the subject. He stated that the canons of syllogistic reasoning were in effect a statement of the symmetrical (De Morgan called it convertible) and transitive character of the relation of identity. He suggested symbols for the converse and the contradictory (he said "contrary") of a relative and for three different ways in which a pair of relatives may be combined (by "relative" he meant what some logicians had called a relative term, i.e. a term which applies to something only in respect of its being related to something else). Those ways can be expressed by the following phrases: (i) x is an l of an m of y (e.g. "John is a lover of a master of Peter"), (ii) x is an l of every m of y, (iii) x is an l of none but an m of y. De Morgan showed that the converse of the contradictory and the contradictory of the converse are identical. He set out in a table the converse, the contradictory and the converse of the contradictory of each of the three combinations and proved that the converse or the contradictory or the converse of the contradictory of each such combination is itself a combination of one of the kinds discussed.

Papers of De Morgan, though containing new and interesting ideas were not easy to read. They were full of ambiguities and technical details which made them difficult to study. Nevertheless they contributed in a significant way to the development of mathematical logic. We shall come back to De Morgan and his works in the next section comparing his achievements with those of Boole.

5.2. G. Boole and his algebra of logic

5.2.1. One of the most important figures in the history of mathematical logic (not only in England) was George Boole (1815–1864). He was interested in logic already in his teens, working as un usher in a private school at Lincoln and educating himself by extensive reading. From that time came his idea that algebraic formulae might be used to express logical relations. His renewed interest in logic was caused by the publication (in periodicals) of letters on the controversy between Sir William Hamilton and Augustus De Morgan on the priority in adoption of the doctrine of the quantification of predicates (cf. the previous section). Boole and De Morgan were in correspondence over a long period (1842–1864).

They exchanged scientific ideas and disucussed various problems of logic and mathematics.

The main logical works of G. Boole are two books: *The Mathematical Analysis of Logic, Being An Essay Toward A Calculus of Deductive Reasoning* (1847) and *An Investigation of the Laws of Thoughts, on which are founded the Mathematical Theories of Logic and Probabilities* (1854).

While working on the pamphlet *Mathematical Analysis of Logic* G. Boole knew already A. De Morgan and they were exchanging letters. De Morgan was working then on his *Formal Logic.* He was meticulous in ensuring that neither could be placed in a position in which he could be accused of plagiarism. He wrote to Boole in a letter of 31st May 1847:

> ... I would much rather not see your investigations till my own are quite finished; which they are not yet for I get something new every day. When my sheets are printed, I will ask for your publication: till then please not to sent it. I expect that we are more likely to have something in common than Sir W.H. (= William Hamilton) and myself.
>
> I should have sent my paper on syllogism ... to you by this post: but I remembered that you might have the same fancy as myself – to complete your own first. Therefore when you choose to have it, let me know (cf. Smith 1982, Letter 11, p.22).

Hence we may conclude that they wrote their works completely independently. There is a story that their books reached the shop on the same day (in November 1847).

The second book of G. Boole *An Investigation of the Laws of Thought* was not very original. It was the result of Boole's studies of works of philosophers on the foundations of logic. The chief novelty of the book was the application of his ideas to the calculus of probabilities. There was no important change on the formal side in comparison with *Mathematical Analysis of Logic.*

We should mention here also his paper "On the calculus of logic" (1848) which contained a short account of his ideas from the pamphlet from 1847. It may be supposed that it was more widely read among mathematicians than *Mathematical Analysis of Logic.*

Boole was not satisfied with the exposition of his ideas in his books. He was working towards the end of his life on the new edition of the *Laws of Thought* preparing various improvements. From the drafts published in (Boole 1952) it follows that what he had in mind was a development of his epistemological views rather than any alteration of the formal side of his work. He mentions in particular a distinction between the logic of class (i.e. his calculus of logic) and a higher, more comprehensive, logic that cannot be reduced to a calculus but may be said to be "the Philosophy of all thought which is expressible in signs, whatever the object of that thought" (cf. Boole 1952, p.14).

5.2.2. Coming now to the detailed analysis of Boole's achievements we must start from the observation that what stimulated his interest in mathematical logic was not only the dispute between De Morgan and Hamilton but also the discussion of the nature of algebra shortly before he wrote his papers (let us mention here the papers of Peacock, Sir William Rowan Hamilton, De Morgan or Gregory – a personal friend of Boole and the editor of "Cambridge Mathematical Journal"). Boole's aesthetic interest in mathematics led him to value very highly all attempts to achieve abstract generality.

Boole started from the following ideas (which were at least implicitly in papers of his contemporaries): (i) there could be an algebra of entities which are not numbers in any ordinary sense, (ii) the laws which hold for types of numbers up and including complex numbers need not all be retained in an algebraic system not applicable to such numbers. He saw that an algebra could be developed as an abstract calculus capable of various interpretations. The view of logic he had can be seen from the opening section of his *Mathematical Analysis of Logic*:

> They who are acquainted with the present state of the theory of Symbolical Algebra, are aware that the validity of the processes of analysis does not depend upon the interpretation of the symbols which are employed, but solely upon the laws of their combination. Every system of interpretation which does not affect the truth of the relations supposed, is equally admissible, and it is thus that the same processes may, under one scheme of interpretation, represent the solution of a question on the properties of numbers, under another, that of a geometrical problem, and under a third, that of a problem of

> dynamics or optics... We might justly assign it as the definitive character of a true Calculus, that it is a method resting upon the employment of Symbols, whose laws of combination are known and general, and whose results admit of a consistent intepretation. That to the existing forms of Analysis a quantitative interpretation is assigned, is the result of circumstances by which those forms were determined, and is not to be construed into a universal condition of Analysis. It is upon the foundation of this general principle, that I purpose to establish the Calculus of Logic, and that I claim for it a place among the acknowledged forms of Mathematical Analysis, regardless that in its objects and in its instruments it must at present stand alone.

What is really new in Boole's works is not the idea of an unquantitative calculus – it was already by Leibniz and Lambert – but a clear description of the essence of the formalism in which the validity of a statement "does not depend upon the interpretation of the symbols which are employed, but solely upon the laws of their combination". Boole indicates here that a given formal language may be interpreted in various ways. Hence he sees logic not as an analysis of abstracts from real thoughts but rather as a formal construction for which one builds afterwords an interpretation. This is really quite new in comparison with the whole tradition including Leibniz. Boole summerizes his position in the last paragraph of the paper "On the calculus of logic" (Boole 1848) as follows:

> The view which these enquires present of the nature of language is a very interesting one. They exhibit it not as a mere collection of signs, but as a system of expressions, the elements of which are subject to the laws of the thougt which they represent. That these laws areas rigorously mathematical as the laws which govern the purely quantitative conceptions of space and time, of number and magnitude, is a conclusion which I do not hesitate to submite to the exactest scrutiny.

One can say that the relation of Boole to Leibniz is the same as was the relation of Aristotle to Plato. In fact, we find by Boole, as it was by Aristotle, not only the ideas, but also a concrete implementation of them, a concrete system. We are going to describe it now.

5.2.3. First of all we must mention that what was built by Boole was not a system in our nowadays meaning – he did not distinguish between properties of operations which are assumed and properties which can be proved or derived from the assumptions.

Symbols x, y, z etc. denoted classes in Boole's papers. He did not distinguish very sharply between class symbols and adjectives. Sometimes he called the letters x, y, z etc. elective symbols thinking of them as symbols which elect (i.e. select) certain things for attention. The symbol $=$ between two class symbols indicated that the classes concerned have the same members. He introduced three operations on classes: (i) intersection of two classes denoted by xy (i.e. the class consisting of all the things which belong to both those classes), (ii) the exclusive union of two classes, i.e. if x and y are two mutually exclusive classes then $x+y$ denotes the class of things which belong either to the class denoted by x or to the class denoted by y, (iii) subtraction of two classes, i.e. if every element of the class y is an element of the class x then $x-y$ denoted the class of those elements of x which are not elements of y. He introduced also special symbols for two classes which form, so to say, limiting cases among all distinguishable classes, namely the universe class, or the class of which everything is a member (denoted by 1) and the null class, or the class of which nothing is a member (denoted by 0). The introduction of those classes involves an interesting novelty with respect to the tradition coming from Aristotle who confined his attention to general terms which were neither universal in the sense of applying to everything nor yet null in the sense of applying to nothing. Boole denoted by 1 what De Morgan called the universe of discourse, i.e. not the totality of all conceivable objects of any kind whatsoever, but rather the whole of some definite category of things which are under discussion.

Having introduced the universe class Boole wrote $1-x$ for the complement of the class x and abbraviated it as $\overline{x}$.

The operation of intersection of classes resembles the operation of multiplication of numbers. But there is an important peculiarity observed by Boole. Namely if x denotes a class then $xx = x$ and in general $x^n = x$. Boole says in *Mathematical Analysis of Logic* that it is the distinguishing feature of his calculus. The analogy between intersection and multiplication suggests also the idea of introducing to the calculus for classes on operation resembling division of two numbers. Boole writes that this analogue may be abstraction. Let x, y, z denote classes related in the way indicated by the equation

$x = yz$. We can write $z = x/y$ expressing the fact that z denotes a class we reach by abstracting from the membership of the class x the restriction of being included in the class y. But this convention has two limits. First the expression x/y can have no meaning at all in the calculus of classes if the class x is not a part of the class y and secondly, the expression x/y is generally indeterminate in the calculus of classes, i.e. there may be many different classes whose intersections with the class y are all coextensive with the class x.

The introduced system of notation suffices to express the A, E, I and O propositions of traditional logic (provided that A and E propositions are taken without existential import). If the symbol x denotes the class of X things and the symbol y the class of Y things, then we have the following scheme:

all X is Y	$x(1-y) = 0$,
no X is Y	$xy = 0$,
some X is Y	$xy \neq 0$,
some X is not Y	$x(1-y) \neq 0$.

Observe that the first two propositions are represented by equalities while the other two by inequalities. Boole prefered to express all the traditional types of categorical propositions by equations and therefore he wrote

some X is Y	as	$xy = v$,
some X is not Y	as	$x(1-y) = v$.

The letter v seems to correspond to the English word "some". It is said to stand for a class "indefinite in all respects but one" – namely that it contains a member or members. But this convention is unsatisfactory. In Boole's system the letter v can be manipulated in some respects as though it were a class symbol. This may suggest mistaken inferences. For example one may be tempted to infer from the equations: $ab = v$ and $cd = v$ that $ab = cd$. Boole himself did not fall into such fallacies by putting restrictions on the use of the letter v which are inconsistent with his own description of it as a class symbol. It seems that a reason and purpose of introducing the symbol v is the desire to construct a device for expressing all Aristotelian inferences, even those depending on existential import.

We find by Boole the following formulae indicating the connections between various operations on classes (those formulae were assumed by him, explicitly or implicitly, as premises):

(1) $xy = yx$,
(2) $x + y = y + x$,
(3) $x(y + z) = xy + xz$,
(4) $x(y - z) = xy - xz$,
(5) if $x = y$ then $xz = yz$,
(6) if $x = y$ then $x + z = y + z$,
(7) if $x = y$ then $x - z = y - z$,
(8) $x(1 - x) = 0$.

The formulae (1)–(7) are similar in form to rules of ordinary numerical algebra. The formula (8) is different. But Boole pointed out that even it can be interpreted numerically. He wrote:

> Let us conceive, then, of an Algebra in which the symbols x, y, z etc. admit indifferently of the values 0 and 1, and of these values alone. The laws, the axioms, and the processes, of such an Algebra will be identical in their whole extent with the laws, the axioms, and the processes of an Algebra of Logic. Differences of interpretation will alone divide them. Upon this principle the method of the following work is established (Boole 1854, pp.37–38).

Those words (and similar in other places) may suggest that Boole's system is a two-valued algebra. But this is a mistake. In fact Boole did not distinguish sharply between his original system (fulfilling rules (1)–(8)) and a narrower system satisfying additionaly the following principle:

(9) either $x = 1$ or $x = 0$.

If we interprete the system in terms of classes then the principle (9) is not satisfied and if we interprete it numerically then the system satisfies even (9).

5.2.4. In both his books *The Mathematical Analysis of Logic* and *An Investigation of the Laws of Thought* Boole suggested a convention that the equations $x = 1$ and $x = 0$ may be taken to mean that the proposition X is true or false, resp. Consequently the truth-values of more complicated propositions can be represented by combinations of small letters, e.g. the truth-value of the conjunction of the propositions X and Y by xy and the truth-value of their exclusive disjunction by $x + y$. This enables us to interprete the Boole's system in terms of the truth-values of propositions with symbols 1 and 0 for truth and falsity respectively. This interpretation was worked out by Boole himself (without use of the phrase "truth-value" which was invented later by G. Frege). He wrote in *The Mathematical Analysis of Logic*:

> "To the symbols X, Y, Z representative of Propositions, we may appropriate the elective symbols x, y, z in the following sense. The hypothetical Universe, 1, shall comprehend all conceivable cases and conjectures of circumstances. The elective symbol x attached to any subject expressive of such cases shall select those cases in which the Proposition X is true, and similarly for Y and Z. If we confine ourselves to the contemplation of a given Proposition X, and hold in abeyance any other consideration, then two cases only are conceivable, viz. first that the given Proposition is true, and secondly that it is false. As these two cases together make up the Universe of Proposition, and as the former is determined by elective symbol x, the latter is determined by the elective symbol $1 - x$. But if the other considerations are admitted, each of these cases will be resoluble into others, individually less extensive, the number of which will depend on the number of foreign considerations admitted. Thus if we associate the Propositions X and Y, the total number of conceivable cases will be found as exhibited in the following scheme.

	Cases	Elective expressions
1st	X true, Y true	xy
2nd	X true, Y false	$x(1-y)$
3rd	X false, Y true	$(1-x)y$
4th	X false, Y false	$(1-x)(1-y)$

> ...And it is to be noted that however few or many those circumstances may be, the sum of the elective expressions representing every conceivable case will be unity (pp.49–50).

In *Laws of Thought* Boole abondoned this interpretation and proposed that the letter x should be taken to stand for the time during which the proposition X is true. This resembles opinions of some ancient and medieval logicians.

5.2.5. In (Boole 1847) and, in a more clear way, in (Boole 1854) we find also another interpretation. In *Mathematical Analysis of*

Logic Boole says that the theory of hypothetical propositions could be treated as part of the theory of probabilities and in *Laws of Thought* we have an interpretation according to which the letter x stands for the probability of the proposition X in relation to all the available information, say K. Hence we have

(10) if X and Y are independent then
$Prob_K(X \text{ and } Y) = xy$,

(11) if X and Y are mutually exclusive then
$Prob_K(X \text{ or } Y) = x + y$.

We see that Boole's system may have various interpretations and Boole himself anticipated that fact. The essence of his method is really the derivation of consequences in abstract fashion, i.e. without regard to interpretation. He made this clear in the opening section of *Mathematical Analysis of Logic.* If he sometimes allows himself to think of his symbols as numerical, that is merely a concession to custom.

5.2.6. The fundamental process in the formal elaboration of Boole's system is so called development. Let $f(x)$ be an expression involving x and possibly other elective symbols and algebraic signs introduced above. Then we have $f(x) = ax + b(1 - x)$. It can be easily seen that $a = f(1)$ and $b = f(0)$. Hence

$$f(x) = f(1)x + f(0)(1 - x).$$

It is said that this formula gives the development of $f(x)$ with respect to x.

If we have an expression $g(x, y)$ with two elective symbols x, y (and the usual symbols of operations), then according to this method we obtain the following development of it:

$$\begin{aligned} g(x, y) &= g(1, y)x + g(0, y)(1 - x), \\ g(1, y) &= g(1, 1)y + g(1, 0)(1 - y), \\ g(0, y) &= g(0, 1)y + g(0, 0)(1 - y), \end{aligned}$$

and consequently using the rule (3)

$$g(x, y) = \\ = g(1, 1)xy + g(1, 0)x(1 - y) + g(0, 1)(1 - x)y + g(0, 0)(1 - x)(1 - y).$$

Factors x, $1-x$ in the development of $f(x)$ and xy, $x(1-y)$, $(1-x)y$, $(1-x)(1-y)$ in the development of $g(x,y)$ are called constituents. We see that for one class x we have 2 constituents, for two classes x, y we have 4 constituents, and in general for n classes we have 2^n constituents. The sum of all constituents of any development is equal 1, their product is equal 0. Hence they may be interpreted as a decomposition of the universe into a pairwise disjoint classes. The development of an expression is treated by Boole (cf. Boole 1847, p.60) as a degenerate case under Maclaurin's theorem about the expansion of functions in ascending powers (he relegates this notion to a footnote in (Boole 1854, p.72)).

5.2.7. The development of elective expressions served Boole to define two other operations, namely solution and elimination. Assume that we have an equation $f(x) = 0$ where $f(x)$ is an expression involving symbol x and maybe other elective symbols. We want to write x in terms of those other symbols. Using development of $f(x)$ one has:

$$f(1)x + f(0)(1-x) = 0,$$

$$[f(1) - f(0)]x + f(0) = 0,$$

$$x = \frac{f(0)}{f(0) - f(1)}.$$

Observe that Boole introduces here the operation of division to which he has assigned no fixed interpretation. But he suggests here a special method of elimination of this difficulty. Developing the expression $\frac{f(0)}{f(0)-f(1)}$ we get a sum of products in which the sign of division appears only in the various coefficients, each of which must have one of the four forms: $\frac{1}{1}, \frac{0}{0}, \frac{0}{1}, \frac{1}{0}$. The first is equal 1 and the third 0. The fourth can be shown not to occur except as coefficient to a product which is separately equal 0, hence it is not very troublesome. The second $\frac{0}{0}$ is curious. Boole says that it is a perfectly indeterminate symbol of quantity, corresponding to "all, some, or none". Sometimes he equates it with the symbol v but it is not quite correct. It is better to use another letter, say w. Hence we have

$$x = \frac{1}{1}p + \frac{0}{0}q + \frac{0}{1}r + \frac{1}{0}s,$$

$$x = p + wq.$$

We can interprete it saying that the class x consists of the class p together with an indeterminate part (all, some, or none) of the class q. In an unpublished draft (cf. Boole 1952, pp.220–226) Boole wrote that the coefficients $\frac{1}{1}, \frac{0}{0}, \frac{0}{1}, \frac{1}{0}$ should be taken to represent respectively the categories of universality, indefiniteness, non-existence and impossibility.

5.2.8. To describe now the rules for elimination assume that an equation $f(x) = 0$ is given. We want to know what relations, if any, hold independently between x and other classes symbolized in $f(x)$. Solving this equation we get

$$x = \frac{f(0)}{f(0) - f(1)}.$$

Hence

$$1 - x = -\frac{f(1)}{f(0) - f(1)}.$$

But $x(1 - x) = 0$, hence

$$-\frac{f(0)f(1)}{[f(0) - f(1)]^2} = 0,$$

i.e.

$$f(0)f(1) = 0.$$

We have now two cases: (i) the working out of the left side of this equation yields $0 = 0$, (ii) it yields something of the form $g(y, z, \ldots) = 0$. In the case (i) we conclude that the original equation covers no relations independent of x, in the case (ii) we establish that the full version of the equation $f(x) = 0$ covers some relation or relations independent of x.

In all those reasonings we see that they involve expressions for which there is no logical interpretation. But Boole has adopted a rule according to which if one is doing some calculations in a formal system then it is not necessary that all formulas in this calculating process have a meaning according to a given interpretation – it is enough that the first and the last ones have such a meaning.

Boole's calculations were sufficient to represent all kinds of reasonings of traditional logic. In particular syllogistic reasonings might be presented as the reduction of two class equations to one, followed by elimination of the middle term and solution for the subject term of the conclusion. Consider, for example, the following reasoning: every human being is an animal, every animal is mortal, hence every human being is mortal. Let h denote the class of human beings, m the class of mortals and a the class of animals. We have by our premises:

$$h(1-a)=0\,,$$
$$a(1-m)=0\,.$$

Hence

$$h-ha+a-am=0\,.$$

Developing this with respect to a we get

$$(h-h1+1-1m)a+(h-h0+0-0m)(1-a)=0,$$
$$(1-m)a+h(1-a)=0\,.$$

Eliminating now a we have

$$(1-m)h=0\,.$$

This means that every human being is mortal.

5.2.9. What was the meaning and significance of Boole's ideas for the development of mathematical logic? First of all we must note that he freed logic from the domination of epistemology and so brought about its revival as an independent science. He showed that logic can be studied without any reference to the processes of our minds. On the other hand Boole's work can be seen as "one of the first important attempts to bring mathematical methods to bear on logic while retaining the basic independence of logic from mathematics" (cf. van Evra 1977, p.374). The chief novelty in his system is the theory of elective functions and their development. Those ideas can be met already by Philo of Megara but Boole was the first who treated these topics in a general fashion. What more, Boole's methods can be applied in a mechanical way giving what is called today a decision procedure. Note also that the symbolism was by him only a tool – we find here no overestimation of the role of a symbolic language. Despite of all imperfections of Boole's methods (e.g. the usage of logically uninterpretable expressions) one must admit that Boole was an important figure in the transitional epoch between the traditional, purely syllogistic logic and the contemporary post-Fregean logic. To give the full account of the development one should mention here also Hugh McColl (1837–1909) who in the paper "The Calculus of Equivalent Statements" (1878–1880) has put forward suggestions for a calculus of propositions in which the asserted principles would be implications rather than equations (as it was by Boole). He proposed an algebraic system of the propositional calculus in the spirit of Boole. This was an attempt to overcome the lacks of Boole's systems and may be treated as a climax in the development of mathematical logic before Frege (who proposed a complete calculus of propositions – cf. the next chapter).

5.2.10. We finish this section with some remarks comparing works and achievements of Boole with those of De Morgan. As we have mentioned above they were in contact over a long time and exchanged scientific ideas. Boole's *Mathematical Analysis of Logic* and De Morgan's *Formal Logic* were written and published simultaneously. Though they worked on their books independently (cf. remarks on this at the beginning of this chapter) their ideas were similar. De Morgan wrote on 27 November 1847 in a draft of a letter to Boole (not sent): "Some of our ideas run so near together, that proof of the physical impossibility of either of us seeing the other's work would be desirable to all those third parties who hold that where plagiarism is possible $1 = a$ whenever a is 0" (cf. Smith 1982, p.24). In a version sent to Boole we find the following words indicating what was the similarity of their approaches: "There are some remarkable similarities between us. Not that I have used the connection of algebraical laws with those of thought, but that I have employed mechanical modes of making transitions, with notation which represents our head work" (cf. Smith 1982, p.25).

Studying their correspondence we come to the conclusion that while in the earlier years the influence flowed from De Morgan to Boole, later the direction has changed. We see nearly always De Morgan reacting to Boole's ideas, rather that vice versa. The influence of De Morgan on Boole was either slight or delayed. G.C. Smith (1982) states explicitly that "it was Boole who had the most original ideas and most vigorous intellect" (p.123). Nevertheless the most important fact was that they both stimulated each other by a steady interaction of ideas.

De Morgan estimated very highly Boole's works. He wrote: "When the ideas thrown out by Mr Boole shall have borne their full fruit, algebra, though only founded on ideas of number in the first instance, will appear like a sectional model of the whole form of thought" (cf. De Morgan 1861, p.346).

5.3. The logical works of Jevons

5.3.1. Boole's works were rather ignored by most contemporary British logicians or damned with faint praise (cf. Gardner 1958). One of persons who saw their importance was William Stanley Jevons (1835–1882), a logician and economist. He regarded Boole's achievements as the greatest advance in the history of logic since Aristotle. But he noticed also defects of Boole's system. He believed that it was a mistake that Boole tried to make his logical notation resemble algebraic notation. "I am quite convinced that Boole's forms (...) have no real analogy to the similar mathematical expressions" – as he stated in a letter (cf. *The Letters and Journals of W. Stanley Jevons* (1886), p.350). He saw also the weakness in Boole's preference for the exclusive rather than the inclusive interpretation of "or".

His own system was supposed to overcame all those defects. Jevons devised a method called by him the "method of indirect inference". He wrote: "I have been able to arrive at exactly the same results as Dr. Boole without the use of any mathematics; and though the very simple process which I am about to describe can hardly be said to be strictly Dr. Boole's logic, it is yet very similar to it and can prove everything that Dr. Boole proved" (cf. Jevons 1870, Lesson 23).

5.3.2. We shall describe now the system of Jevons. He introduced the following operations performed on classes: intersection, complement and the inclusive union $+$ (denoted by him originally by $\cdot|\cdot$ to distinguish it from Boole's exclusive union). The null class was denoted by 0 and the universe class by 1. The expression $X = Y$ indicated the equality of the classes X and Y, i.e. the fact that they have the same elements. To express the inclusion of X in Y

Jevons wrote $X = XY$. He noted the following properties of those operations:

$$\begin{aligned}&XY = YX\,,\\&X + Y = Y + X\,,\\&X(YZ) = (XY)Z\,,\\&X + (Y + Z) = (X + Y) + Z\,,\\&X(Y + Z) = XY + XZ\,.\end{aligned}$$

He used also the principle of identity $X = X$, of contradiction $Xx = 0$ (Jevons symbolized a complement using a lower-case letter – he borrowed this convention from De Morgan, cf. section 1) as well as *tertium non datur* $X + x = 1$ and the rule of substitution of similars. The last rule is used by him to verification of inferences. Take for example the following reasoning (known in the traditional logic as the mood Barbara):

$$\begin{array}{cccc}\text{All} & S & \text{is} & M\\ \text{All} & M & \text{is} & P\\ \hline \text{All} & S & \text{is} & P\end{array}$$

or symbolically

$$\begin{array}{c}SaM\\ MaP\\ \hline SaP\end{array}$$

The first premise is written by Jevons as $S = SM$, the second as $M = MP$. Substituting M for MP in the first premise we get $S = S(MP) = (SM)P$. But $S = SM$, hence $S = SP$, i.e. SaP.

Jevons developed his logic using properties of operations and Boole's idea of constituents. The easiest way to explain it is to give an example. Consider the following syllogistic premises: "All A is B" and "No B is C". First we write a table exhausting all posible combinations of A, B, C and their negations (cf. Boole's constituents):

$$\begin{array}{l}ABC\\ ABc\\ AbC\\ Abc\\ aBC\\ aBc\\ abC\\ abc\,.\end{array}$$

Jevons called such a list an "abecedarium" and later a "logical alphabet". The premise "All A is B" tells us that the classes Abc and AbC are empty, similarly from "No B is C" it follows that ABC and aBC are empty. The final step is now to inspect the remaining classes, all consistent with the premises, to see what we can determine about the relation of A to C. We note that "No A is C" (there are no remaining combinations containing both A and C). In this way we got a reasoning known in the traditional logic as the mood Celarent. Similarly, under the additional assumption that none of the three classes A, B, C is empty, we can conclude that "Some A is not C", "Some not-A is C" etc. What is important here is the fact that Jevons's system was not limited to the traditional syllogistic conclusions (and he was very proud of it).

Jevons, as most of other logicians of that time, confined his attention almost exclusively to class logic. He combined statements of class inclusion with conjunctive or disjunctive assertions but almost never worked with truth-value relations alone. He prefered also, like Boole, to keep his notation in the form of equations (writing for example $A = AB$ to express the fact that all A is B). This preference for "equational logic" – as Jevons called it – was one of the reasons of the fact that he did not consider problems of a truth-value nature (though his method operates very efficiently with such problems).

5.3.3. The methods of Boole and of Jevons were suitable for a mechanization. Jevons devised a number of laborsaving devices to increase the efficiency of his method. The most important were his "logical abacus" and his "logical machine". The first, described in Jevons (1869) "consisted of small rectangular wooden boards, all the same size, and each bearing a different combination of true and false terms. The boards were lined up on a rack. An arrangement of pegs on the side of each board was such that one could insert a ruler under the pegs and quickly pick out whatever group of boards one wished to remove from the rack" (Gardner 1958).

Jevons's logical machine was described by him in (Jevons 1870) and in (Jevons 1874). It was built for him by a young clockmaker in Salford in 1869. In 1870 Jevons demonstrated his machine at a meeting of the Royal Society of London. In 1914 his son gave it to the Science Museum, South Kensington, London but in 1934 it was transferred to the Oxford Museum of the History of Science, Old Ashmolean Building, Oxford, where it is now on display.

Jevons's logical machine resembles a miniature upright piano about 3 feet high. On the face of it are openings through which one can see letters representing the 16 possible combinations of four terms and their negations (the constituents). Each combination forms a vertical row of four terms. The keyboard consists of 21 keys arranged in the following way:

<table><tr><td>FINIS</td><td>·|·</td><td>d</td><td>D</td><td>c</td><td>C</td><td>b</td><td>B</td><td>a</td><td>A</td><td>COPULA</td><td>A</td><td>a</td><td>B</td><td>b</td><td>C</td><td>c</td><td>D</td><td>d</td><td>·|·</td><td>FULL STOP</td></tr></table>

We have 5 "operational keys" and 16 keys representing terms, positive and negative. The "copula" is pressed to indicate the sign of equality connecting left and right sides of an equation. The "full stop" is pressed after a complete equation has been fed to the machine. The "finis" key restores the machine to its original condition. The keys "·|·" represent the inclusive "or". They are used whenever the "or" relation occurs within either the left or right sides of an equation.

The machine is operated by pressing keys in the order indicated by the terms of an equation. If we have for example an equation $A = AB$ (i.e. "All A is B") then we press keys in the following order: A (on the left), copula, A (on the right), B (on the right), full stop. This action automatically eliminates from the face of the machine all combinations of terms inconsistent with the proposition just fed to the machine. Additional equations are handled exactly in the same manner. After all premises have been fed to the machine, its face is then inspected to determine what conclusions can be drawn.

Jevonns' machine had no practical use. One of the reasons was the fact that complex logical questions seldom arise in everyday life. It could be used, say, as a classroom device for demonstrating the nature of logical analysis. The machine itself had also several defects (what is quite understandable for it was the first of its kind): (1) since statements should be fed to the machine in a clumsy equational form, it is made unnecessarily complicated, (2) there was no efficient procedure of feeding "some" propositions to it, (3) it was imposible to extend the mechanism to a large number of terms (Jevons planed to build a machine for ten terms but abondoned the project when it became clear that the device would occupy the entire wall space of one side of his study), (4) the machine merely exhibits all the consistent lines of the logical alphabet but it does not performe the analysis of them to indicate the desired conclusion.

5.3.4. The Jevons's idea of building a logical machine was developed after him. Several other devices were proposed. We want to mention here only Allan Marquand (1853–1924) and his machine from 1881–1882, Charles P. R. Macaulay whose machine (built in 1910) combined the best features of both Jevons's and Marquand's machines being an extremely compact and easily operated device and Annibale Pastore (b. 1868) and his machine constructed in 1903. They were all mechanical devices and ancestors of modern electrical machines. The first electrical analogue of Jevons's machine was suggested by A. Marquand in 1885 and in 1947 an electrical computer was constructed at Harvard by T.A. Kalin and W. Burkhart for the solution of Boolean problems involving up to twelve logical variables (i.e. propositions or class letters).

5.3.5. What was the import of Jevons's contribution to the development of logic? First of all he developed the ideas and methods of Boole removing some of their defects, e.g. introducing the inclusive interpretation of "or" what made possible a great simplification and what was welcome by all later writers on the algebra of logic. He did also very much to mechanize Boole's methods inventing various devices which facilitated the practical usage of them.

5.4. J. Venn and logical diagrams

5.4.1. Talking about the mechanization of Boole's methods and about Jevons we must pay some attention to John Venn (1834–1923) and to the idea of logical diagrams. Venn did not have a high opinion of Jevons as a logician. He wrote (1881): "Excellent as much of Jevons's work is, I cannot but hold that in the domain of logic his inconsistencies and contradictions are remarkable". There was a strong rivalry between the two men. Venn dismissed Jevons's logic machine as essentially trivial. He developed an idea of logical diagrams which were, in his opinion, a better and more efficient device to solve logical problems.

5.4.2. Before discussing Venn's diagrams in detail we say some words about his logic in general.

In Venn's opinion the aim of symbolic logic was to build a special language which would enable "to extend possibilities of applying our logical processes using symbols" (cf. Venn 1881, p.2).

He used Latin letters as symbols of classes, 0 and 1 as symbols of the null and universe classes, resp., introduced complement, intersection, union, subtraction and division of classes. By $x - y$ he meant the class remaining after elimination from the class x the elements of the class y provided that y is a part of x. Subtraction was a converse operation to the union. Similarly division was considered as a converse to the intersection. It was not determined uniquely. The expression $x > 0$ denoted the fact that the class x is not empty. It was a negation of the expression $x = 0$. The equation $x = y$ was understood by Venn as the sentence "there are no entities which belong to x but do not belong to y and there are no entities belonging to y but not to x". He wrote it as $x\overline{y} + \overline{x}y = 0$. Venn formulated also several properties of operations (not proving them at all).

5.4.3. As one of the main aims of symbolic logic Venn considered solution of logical equations and elimination of variables. Contrary to Boole and Jevons he considered not only equations but also inequalities. Solving such problems he used algebraical methods as well as diagrams. The latter is one of his main achievements in symbolic logic and we want now to study it carefully.

One can define a logical diagram as a two-dimensional geometric figure with spatial relations that are isomorphic with the structure of a logical statement. Historically the first logic diagrams probably expressed statements in what today is called the logic of relations. The tree figure was certainly known to Aristotle, in medieval and Renaissance logic one finds the so called tree of Porphyry which is another example of the type of diagram. We meet trees also by Raymond Lull, we see diagrams in various "squares of opposition" (showing certain relations of immediate inference from one class proposition to another), in "pons asinorum" of Petrus Tartaretus.

As a first important step toward a diagrammatic method sufficiently iconic to be serviceable as a tool for solving problems of class logic we may consider the use of a simple closed curve to represent a class. It is impossible to say who used it as first. We must mention here J. Ch. Sturm and his *Universalia Euclidea* (1661), G. W. Leibniz (who used circles and linear diagrams), J. Ch. Lange (and his *Nucleus logicae Weisianae* 1712), J. H. Lambert (and his linear method of diagramming explained in *Neues Organon* 1764) and L. Euler. The latter introduced them into the history of logical analysis (cf. Euler 1768–1772).

J. Venn's method is a modification of Euler's. It has elegantly overcome all limitations of the latter. Venn first published his method in an article from 1880 and then discussed it more fully in his book from 1881.

To describe the Venn's method let us show first how does it apply to a syllogism. Consider for example the following reasoning (the mood Celarent):

$$\begin{array}{c} \text{All } S \text{ is } M \\ \text{No } M \text{ is } P \\ \hline \text{No } S \text{ is } P. \end{array}$$

Draw three circles intersecting each other and representing, resp., S (subject), M (middle term), P (predicate). All the points inside a given circle are regarded as members of the class represented by the given circle, all points outside it – as members of the complement of this class. We get the following picture:

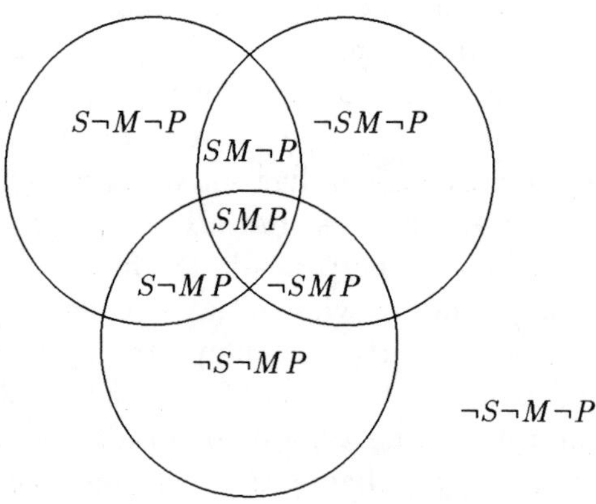

It represents simply the decomposition of the universe (represented by the plane) into 8 constituents (we use the symbol $\neg$ to denote the negation). If we wish now to show that a given compartment is empty we shade it. If we wish to show that it has members, we place a small x inside it. If we do not know whether an element belongs to one compartment or to an adjacent one, we put x on the border between the two areas. Hence to diagram our first premise "All S is M" we must shade all compartments in which we find S and $\neg M$ (because we interprete it to mean the class of things which are S and $\neg M$ is empty), i.e. compartments $S\neg M\neg P$ and $S\neg MP$. The second premise "No M is P" says that all compartments containing MP are empty. Hence we shade compartments SMP and $\neg SMP$. We get the following diagram:

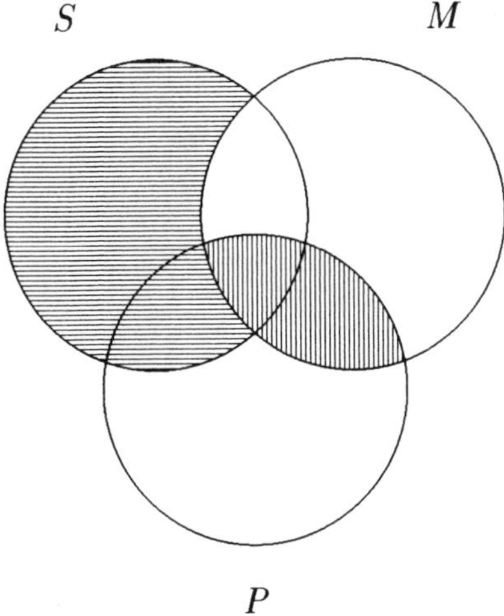

One sees that all areas containing both S and P are empty; hence we may conclude "No S is P". Consequently our scheme is correct. If we assume that S is not an empty class, we may also conclude that "Some S is not P".

Simply modification of Venn's diagram allow us to take care of numerical syllogism in which terms are quantified by "most" or by numbers. We must change at least one of the circles to a rectangle. Here is an example indicating how one can diagram the syllogism:

there are ten A's of which four are B's; eight A's are C's; therefore at least two B's are C's.

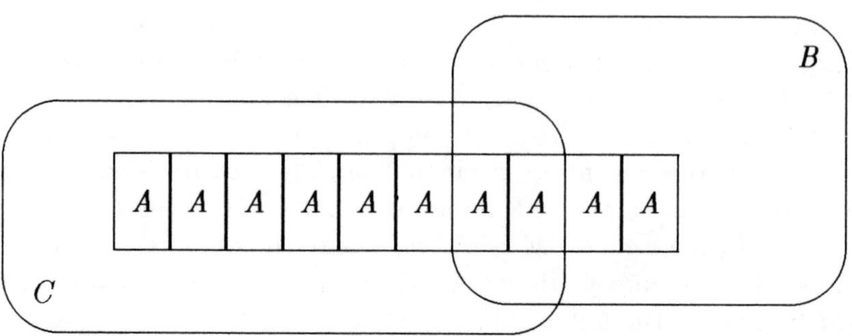

One of the merits of Venn's system of diagrams is that it can be extended in principle to take care of any number of terms, but, of course, as the number of terms increases, the diagram becomes more involved. For four terms we use ellipses to diagram all the constituents:

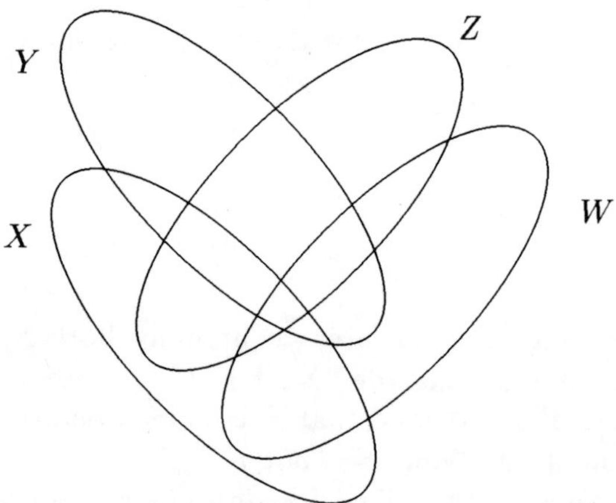

For five terms Venn proposed the following diagram:

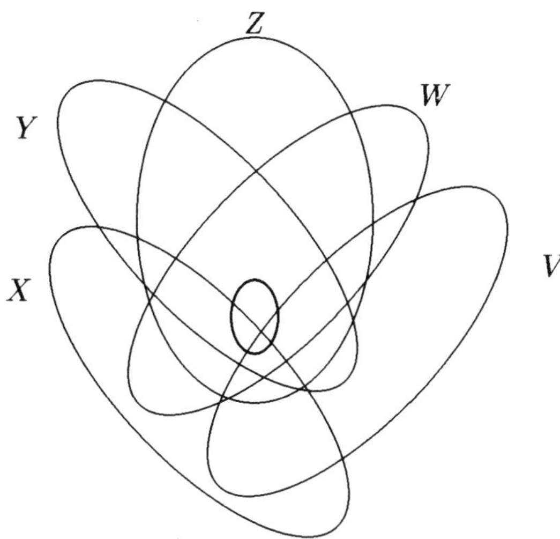

The diagram has a defect that the class Z has a shape of a doughnut – the small ellipse in the center being outside Z but inside W and Y. Beyond five terms Venn proposed to divide a rectangle into the desired number of subcompartments, labeling each with a different combination of the terms.

5.4.4. The method of Venn's diagrams has been developed and improved. We mention here only diagrams of Allan Marquand (lying on the border line between a highly iconic system of Venn and a noniconic system of notation) (cf. Marquand 1881), Alexander Macfarlane (cf. Macfarlane 1885, 1890), W.J. Newlin (and his method of deviding a square) (cf. Newlin 1906), W.S. Hocking (cf. Hocking 1909) and diagrams of an English logician and mathematician Lewis Carol (whose proper name was Dodgson Charles Lutwidge; 1832–1898). He explained his method in two books: *The Game of Logic* (1886) and *Symbolic Logic* (1896). It was similar to Marquand's divisions of a square – apparently Carrol was not familiar with Marquand's work. Carrol's method easily took care of syllogisms with mixtures of positive and negative forms of the same term, it could be extended to n terms (in *Symbolic Logic*

Carrol pictured a number of these extensions including a 256-cell graph for eight terms).

5.4.5. Venn's diagrams can be used for solving problems not only in the class logic but also in the propositional calculus. We must only interprete them in a different way: each circle stands now for a proposition which may be either true or false, the labels in the various compartments indicate possible or impossible combinations of true and false values of the respective terms, we shade a compartment to indicate that it is an impossible combination of truth values (an unshaded compartment indicates a permissible combination). Using this procedure we can solve with the help of diagrams various problems of propositional logic which concern simple statements without parentheses. In the latter case some insolvable difficulties arise.

5.4.6. Nevertheless diagrams developed by J. Venn (and later improved by others) proved to be a fruitful and useful method in the algebra of logic. It is not true that J. Venn only improved slightly Euler's method by changing shapes of figures. The main difference between Euler and Venn lies in the fact that Venn's method was based on the idea of decomposition of the universe into constituents (which we do not have by Euler). Diagrams served Venn not only to illustrate solutions obtained in another way but they were in fact a method of solving logical problems.

5.5. Conclusions

Having presented ideas and works of English logicians of 19th century we come now to conclusions. What was the meaning and significance of their achievements, what was their contribution to the development of mathematical logic?

Mathematical logic of 19th century was developed in connection with mathematics, i.e. mathematics was a source of patterns, a model for logic, served as an ideal which was imitated. Mathematical methods – first of all methods of formal algebra – were applied to logic. Therefore the main form of logic in 19th century was the algebra of logic. The analogy between logic and algebra which led to the origin of the algebra of logic was the following: a solution of any problem by solving an appropriate equation is in fact a derivation of certain consequences from the initial conditions of the problem. Hence the idea of extending this method to problems of unquantitative character. This tendency had a source and a good motivation also in Leibniz's works and ideas (cf. his project *characteristica universalis*).

In this way, starting from the traditional syllogistic logic and developing ideas of Leibniz and applying some ideas of formal algebra, the logicians of the 19th century came to the idea of the algebra of logic. Those ideas were developed first of all just by English logicians. De Morgan may be treated here as a forerunner and Boole was the man who established the main principles and tools. The problem was to develop, in analogy to the algebra, a method of expressing unquantitative information in the form of equations and to establish rules of transformation of those equations. This led to the construction of systems called today Boolean algebras. They were formulated in the language of classes as extension of concepts. Later another interpretation – namely the logic of propositions – was developed. In this way logic was freed from the domination of epistemology and so its revival as an independent science was brought. It was shown that logic can be studied without any reference to the processes of our mind.

The idea of the algebraization of logic led also to the search for mechanical methods of solving logical problems. Various methods based on the usage of logical machines and diagrams were proposed (Jevons, Venn, Carrol and others). Considerations of Boole and De Morgan led also to the development of the general theory of relations.

We may say that English logicians played first fiddle in mathematical logic in the 19th century. Of course their contributions did not contain the ideas in a perfect and final form. They had various drawbacks and defects. Later logicians developed them removing the imperfections (let us mention here works of Peirce, Gergonne and Schröder). Nevertheless the works of English logicians formed the main step in the transition epoch between the traditional syllogistic logic and the prehistory of mathematical logic on the one hand and the Fregean period on the other. In the former the very idea of mathematical logic arose, first of all due to Leibniz. The latter was characterized by new aims of logic – it was then developed to build foundations of mathematics as a science, mathematics and mathematical methods were the subject of logical studies. The transitional period, called in the literature the Boolean (!) period (cf. Bocheński 1956) was necessary to create mathematical logic as an independent science. It was done by using mathematics as a pattern, using mathematical, in particular algebraic, methods. This period was started with De Morgan's *Formal Logic* and Boole's *The Mathematical Analysis of Logic* (1847) and ended with Schröder's *Vorlesungen über die Algebra der Logik* (1890–1905). Works of English logicians pointing out the directions of research and establishing the fundamental ideas and methods were crucial in this period.

CHAPTER SIX

THE 20TH CENTURY WAY TO FORMALIZATION AND MECHANIZATION

6.1. Introduction

The aim of this chapter is to discuss the development of the idea of formalization and mechanization in the 20th century. The logic of this time grew from two logical traditions which had arisen during the latter half of the 19th century. The first, often called algebraic, originated with A. De Morgan and G. Boole and grew through the researches of their English followers such as W. S. Jevons and J. Venn (cf. Chapter 5) as well as those of Ch. S. Peirce in the United States and E. Schröder in Germany. They treated the analogy between the laws of algebra and the laws of logic as a guide in their logical investigations. Their aim was to develop a method of expressing unquantitative information in the form of equations and to establish rules of transformations of those equations. Systems constructed by them have various interpretations: class-theoretic, propositional etc. Generally speaking they investigated the following sort of questions: given an equation between two expressions of the calculus, can that equation be satisfied in various domains. It should be admitted that they had no notion of a formal proof.

The second tradition, often called the logistic method, can be characterized by a new aim – it was developed to build foundations of mathematics as science and consequently mathematics and mathematical methods were no longer treated as a pattern for logic but rather as a subject of logical studies. G. Peano and G. Frege can be seen as founders of this tradition. It was later developed and perfected by B. Russell.

Since the algebraic tradition was rather a transition from the traditional syllogistic logic to modern logic and the latter was influenced mainly by the logistic tradition, hence we shall start our considerations with the discussion of G. Peano's and G. Frege's achievements. Next the work of Russell will be presented. Further we shall discuss achievements of T. Skolem, D. Hilbert and his school, the Gentzen's idea of natural deduction, works of Herbrand and we shall finish with the discussion of semantic tableaux of Beth.

6.2. Peano's symbolism and axioms for mathematics

6.2.1. Early logical works of the Italian mathematician Giuseppe Peano (1858–1932) were based on ideas of the algebra of logic. One can see it for example in his first logical work – in the 20 page introductory chapter to the book *Calcolo geometrico secondo l'Ausdehnungslehre di H. Grassmann, precedutto dalle operationi della logica deduttiva* (1888). Peano used there the logic of G. Boole, Ch. S. Peirce and E. Schröder in mathematical investigations but introduced into it a number of innovations: he used for example different signs for logical and mathematical operations, introduced the distinction between categorical and conditional propositions that was to lead him to quantification theory.

6.2.2. Already in this paper one can observe that Peano set great store by mathematical symbolism. The propagation of a symbolic language in mathematics and the invention of a good symbolism is just one of his greatest merits. Peano's activity in this direction was strongly connected with the idea of *characteristica universalis* of Leibniz. He considered the realization of Leibniz's ideas as a main task of the mathematical sciences.

Peano developed symbolic language to use it later for axiomatization of mathematics. The first attempt in this direction was his small book *Arithmetices principia nova methodo exposita* (1889). This book with a Latin title (to be precise we should say: almost Latin – the word 'arithmetices' is only a transliteration of an appropriate Greek word, in Latin it ought to be 'arithmeticae'), written also in Latin contains the first formulation of Peano's postulates for natural numbers.

The work consists of a preface and a proper part. In the preface needed notions are introduced and a symbolism is fixed. G. Peano wrote at the beginning:

> I have represented by signs all ideas which appear in the foundations of arithmetic. With the help of this representation all

> sentences are expressed by signs. Those signs belong either to logic or to arithmetic. ... By this notation every sentence become a shape as precise as an equation in algebra. Having sentences written in such a way we can deduce from them other sentences – it is done by a process similar to a looking for a solution of an algebraic equation. This was exactly the aim and the reason for writing this paper.

The logical part of the work presents formulas of the propositional calculus, of the calculus of classes and a few of the quantification theory. The notation used is much better than that of Boole and Schröder. G. Peano distinguishes between the propositional calculus and the algebra of classes. A new and useful notation for the universal quantifier is introduced: if a and b are two formulas containing free variables $x, y, \ldots$ then Peano writes $a \supset_{x,y,\ldots} b$ to denote what we would write today as $\forall x \forall y \ldots (a \longrightarrow b)$.

The aim of the proper part of the book is to rewrite arithmetic in symbolic notation. The primitive arithmetical notions are: number (N), one (1), successor ($x+1$), and equality ($=$). Peano formulates four identity axioms (treated today as logical ones) and five proper arithmetical axioms. In his notation they look like this:

$$
\begin{aligned}
&1 \,\varepsilon\, N, \\
&a \,\varepsilon\, N . \supset .a + 1 \,\varepsilon\, N, \\
&a, b \,\varepsilon\, N . \supset : a = b . = .a + 1 = b + 1, \\
&a \,\varepsilon\, N . \supset .a + 1 - = 1, \\
&k \,\varepsilon\, K .\!\cdot\!. 1 \,\varepsilon\, k .\!\cdot\!. x \,\varepsilon\, N . x \,\varepsilon\, k :\supset_x .x + 1 \,\varepsilon\, k :: \supset .N \supset k,
\end{aligned}
$$

where K denotes the family of all classes, ε is the membership relation, $\supset$ denotes inclusion and simultaneously implication. In the notation used today they would be written as follows:

$$
\begin{aligned}
&1 \in N, \\
&a \in N \longrightarrow a + 1 \in N, \\
&a, b \in N \longrightarrow (a = b \equiv a + 1 = b + 1), \\
&a \in N \longrightarrow (a + 1 \neq 1), \\
&[k \in K \wedge 1 \in k \wedge \forall x (x \in N \wedge x \in k \longrightarrow x + 1 \in k)] \longrightarrow N \subset k.
\end{aligned}
$$

In the sequel Peano develops on the basis of these axioms not only the arithmetic of natural numbers but deals also with fractions, real numbers, even with the notion of limit and definitions in point-set theory.

Peano introduced new notions by definitions. They were in fact recursive definitions not satisfying the general conditions he put on definitions. He claimed namely that a definition must be of the form $x = a$ or $\alpha \supset x = a$ where α is a certain condition, x is a notion being defined and a is "an aggregate of signs having a known meaning". He formulated no theorem (similar to Theorem 126 from the paper of Dedekind 1888) which would justify definitions by recursion.

But it is not the only defect of Peano's *Principia.* A more grave one is the fact that formulas are simply listed, not derived – and they could not be derived, because no rules of inference were given and no notion of a formal proof was introduced. This is perhaps a consequence of the fact that Peano was interested in logic only as in a tool which helps us to order the mathematics and to make it more clear. He was not interested in logic for itself.

What is presented by Peano as a proof is actually a list of formulas which are such that, from the point of view of a working mathematician, each formula is very close to the next. But one cannot speak here about derivation or deduction. This brings out the whole difference between axiomatization and formalization.

The absence by Peano of, for example, the modus ponens is connected with his inadequate interpetation of the conditional. He read '$a \supset b$' as 'from a one deduces b' ('ab a deducitur b') which remains vague. He did not use truth values at all.

In a series of papers (cf. Peano 1891, 1891a, 1891b) Peano undertook to prove the logical formulas that he simply listed in *Principia.* But again those proofs suffer from the absence of rules of inference.

Some of Peano's remarks could suggest that his logical laws should be taken as rules of inference, not as formulas in a logical language. But this would not yield a coherent interpretation of his system.

6.2.3. At the end of the introduction to *Arithmetices principia* Peano wrote: "My book ought to be considered as a sample of this new method. Using the notation which we have introduced one can formulate and prove infinitely many other theorems, in particular theorems about rational and irrational numbers". He was convinced that with the help of the introduced symbols, perhaps supplemented by some new symbols denoting new primitive notions, one can express in a clear and precise way theorems of every science.

In his succeeding papers G. Peano successively showed how to apply this method to mathematics. He wrote various papers reducing to a minimum the usage of colloquial language. One must mention here the monograph *I principii de geometria logicamente esposti* (Torino 1889, pp. 40) – published in the same year as *Arithmetices principia.* He used here logical and arithmetical symbols and some additional special geometrical symbols. His system of geometry is based on two primitive notions, namely on the notion of point and interval. It is similar to the approach of Pasch. Geometry was just the second domain of mathematics where Peano successfully applied his symbolic-axiomatic method. Investigating the foundations of geometry G. Peano popularized the vector method of H. Grassmann and brought it to perfection. In this way he also gave a certain impulse to the Italian school of vector analysis.

The idea of presenting mathematics in the framework of an artificial symbolic language and of deriving then all theorems from some fundamental axioms was also the basic idea of Peano's famous project *Formulario.* It was proclaimed in 1892 in the journal *Rivista di Matematica.* This journal was founded by Peano in 1892 and here his main logical papers and papers of his students were published. The aim of his project was to publish all known mathematical theorems. This plan was to be carried out using, of course, the symbolic language introduced by Peano. A special journal *Formulario Mathematico* was founded to publish the results of this project. It was edited by Peano and his collaborators: Vailati, Castellano, Burali-Forti, Giudice, Vivanti, Betazzi, Fano and others. Five volumes appeared: Introduction 1894, vol. I–1895, vol. II–1897–1899, vol. III–1901, vol. IV–1903, vol. V–1908; this last volume contained about 4200 theorems! Peano even bought a small printing office (in 1898 from Faà di Bruno for 407 lira) and studied the art of printing. One of the results of the project was a further simplification of the mathematical symbolism. Peano treated *Formulario* as a fulfilment of Leibniz's ideas. He wrote: "After two centuries, this 'dream' of the inventor of the infinitesimal calculus has become a reality... We now have the solution to the problem proposed by Leibniz". He hoped that *Formulario* would be widely used by professors and students. He himself immediately began to use it in his classes. But the response was very weak. One reason was the fact that almost everything was written there in symbols, the other was the usage of *latino sine flexione* (that is Latin without inflection – an artificial international language invented by Peano – for details cf. e.g. (Murawski 1985) or (Murawski 1987)) for explanations and commentaries (instead of, for example, French).

6.2.4. What was the real significance of Peano's writings for the formalization and mechanization of reasonings, what were his real merits and achievements? One should admit that his works were of minor significance for logic proper, but they showed how mathematical theories can be expressed in one symbolic language. Peano believed that this would help to answer the question: "how to recognize a valid mathematical proof?" He wrote in *I principii di geometria logicamente esposti:*

> ... this question can be given an entirely satisfactory solution. In fact, reducing the propositions ... to formulas analogous to algebraic equations and then examining the usual proofs, we discover that these consist in transformations of propositions and groups of propositions, having a high degree of analogy with the transformation of simultaneous equations. These transformations, or logical identities, of which we make constant use in our argument, can be stated and studied.

The influence of Peano and his writings on the development of mathematical logic was great. In the last decade of the 19th century he and his school played first fiddle. Later the center was moved to England where B. Russell was active – but Peano has played also here an important role. Russell met Peano at the International Philosophical Congress in Paris in August 1900 and later he wrote about this meeting in the following way:

> The Congress was a turning point in my intellectual life, because I met there Peano. I already knew him by name and had seen some of his work, but had not taken the trouble to master his notation. In discussions at the Congress I observed that he was always more precise than anyone else, and that he invariably got the better of any argument upon which he embarked. As the days went by, I decided that this must be

> owing to his mathematical logic. I therefore got him to give me all his works, and as soon as the Congress was over I retired to Fernhurst to study quietly every word written by him and his disciples. It became clear to me that his notation afforded an instrument of logical analysis such as I had been seeking for years, and that by studing him I was acquiring a new and powerful technique for the work I had long wanted to do. (cf. Russell 1967).

In a letter to Jourdain from 1912 B. Russell wrote: "Until I got hold of Peano, it had never struck me that Symbolic Logic would be any use for the Principles of mathematics, because I knew the Boolean stuff and found it useless. It was Peano's, together with the discovery that relations could be fitted into his system, that led me to adopt symbolic logic." It should be admitted also that Russell learned of Frege and his works just through Peano (cf. Kennedy 1973).

Thanks to those facts the writings of Peano rapidly gained a wide influence and greatly contributed to the diffusion of the new ideas.

6.3. G. Frege and the idea of a formal system

6.3.1. Contemporarily with G. Peano lived the German mathematician and logician Gottlob Frege (1848–1925). Frege's system ushered in the modern epoch in the history of logic. The year 1879 – the year of the publication of Frege's *Begriffsschrift* is considered as the beginning of modern logic. As J. van Heijenoort writes, Frege "freed logic from an artificial connection with mathematics but at the same time prepared a deeper interrelation between these two sciences" (cf. Heijenoort 1967, p.2).

G. Boole and the whole algebraic tradition wished to exhibit logic as a part of mathematics. Frege wanted to realize a different aim which seemed to be inconsistent with the former – namely he wished to show that arithmetic, and consequently mathematics (the so-called arithmetization of analysis has already shown that analysis and in particular the theory of real and complex numbers can be reduced to arithmetic of natural numbers) is identical with logic. There is in fact no inconsistency between these two tendencies. When Boole wrote of the mathematical analysis of logic he meant only the presentation of logic as a calculus similar in certain respects to numerical algebra (cf. Chapter 4). Frege went much further. He not only demanded a formal rigour within the study of logic but tried also to show that numbers can be defined without reference to any notions other than those involved in the interpretation of his calculus as a system of logic. To do it he had to construct a precise system including the traditional material and the newer contributions of Leibniz and Boole, a system being more regular than ordinary language and having the property that everything that is required for the proof of theorems is set out explicitly at the beginning and the procedure of deduction is reduced to a small number of standard moves (to avoid the danger of smuggling in what one ought to prove). Such a system was constructed by Frege in the little book (88 pages) *Begriffsschrift, eine der arithmetischen nachgebildete Formelsprache des reinen Denkens* (1879). This was Frege's first work in the field of logic.

6.3.2. Wanting to free logic from too close attachment to the grammar of ordinary language, Frege devised a language that deals with the *conceptual content* which he called *Begriffsschrift.* This term is usually translated into English as 'ideography' (a term used in Jourdain (1912) – this paper was read and annotated by Frege) or, as Austin has proposed, as 'concept writing' (cf. Frege 1950).

Frege wrote about his language (in an unpublished fragment dated 26 July 1919): "I do not start from concepts in order to build up thoughts or propositions out of them; rather, I obtain the components of a thought by decomposition (*Zerfällung*) of the thought. In this respect my Begriffsschrift differs from the similar creations of Leibniz and his successors – in spite of its name, which perhaps I did not chose very aptly". Answering to the censure for wrecking the tradition established in the previous thirty years (cf. e.g. Schröder 1880) he wrote in 1882:

> My intention was not to represent an abstract logic in formulas, but to express a content through written signs in a more precise and clear way than it is possible to do through words. In fact, what I wanted to create was not a mere *calculus ratiocinator* but a *lingua characteristica* in Leibniz's sense (Frege 1882).

Frege's ideography is a 'formula language', a language written with special symbols, 'for pure thought', that are manipulated according to definite rules. Wanting to construct a system of logic which would provide a foundation for arithmetic he could employ neither the algebraical symbolism nor the analogies between logic and arithmetic uncovered by Boole and others. Though the subtitle of Frege's book says that his ideography is to be modelled on the language of arithmetic still the main resemblence is in fact in the use of letters to express generality.

One of the great innovations of *Begriffsschrift* is the analysis of a sentence not in terms of subject and predicate, as it was so far, but in terms of function and arguments. This idea, coming from mathematics, had a profound influence upon modern logic.

According to Frege, all symbols have meaning. The meaning of a sign which is a subject in the linguistical sense, is an object. In a symbolic language every expression must denote an object and one cannot introduce any symbol without meaning. Expressions of such a language denote either objects or functions. The word 'object' (*Gegenstand*) must be understood here in a narrower sense – it denotes a described individual, a class or the logical value (which means here 'truth' (*das Wahre*) and 'falsity' (*das Falsche*). But sometimes Frege used the word 'object' more generally, speaking about 'saturated objects' (i.e. objects in a proper sense) and about 'unsaturated objects' (i.e. functions). By calling functions 'unsaturated objects' he meant that they are not completely defined. They have one or more unsaturated places. In particular, functions with one unsaturated place whose values are truth values he called concepts, and functions with more unsaturated places – relations. Those functions were for him timeless and spaceless ideal objects.

6.3.3. To describe the symbolism of Frege notice first of all that it was two-dimensional which was a departure from the usual practice – previously people had expressed their thoughts in a linear one-dimensional form.

Frege begins by introducing a sign of content and a sign of judgement. If A is a sign or a combination of signs indicating the content of the judgement then the expression $—A$ means: 'the circumstance that A' or 'the proposition that A', and the symbol $\vdash A$ expresses just the judgement that A. Frege wrote in *Begriffsschrift:*

> Let ⊢ A stand for (*bedeutet*) the judgement 'opposite magnetic poles attract each other'; then —A will not express (*ausdrücken*) this judgement; it is to produce in the reader merely the idea of the mutual attraction of opposite magnetic poles, say in order to derive consequences from it and to test by means of these whether the thought is correct.

Hence the sign ⊢ is the common predicate for all judgements.

The expression

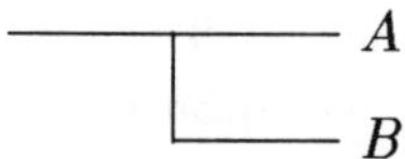

means: 'the circumstance that the possibility that A is denied and B is affirmed does not take place' (the vertical line linking horizontals was called here 'condition stroke'). If we add to it the vertical stroke then we get

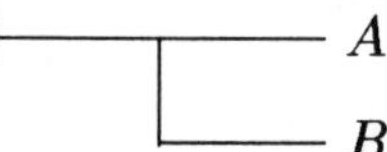

which stands for the judgement 'the possibility that A is denied and B affirmed does not take place'. Hence it is a symbol of the Philonian conditional with B as antecedent and A as consequent. We see that Frege's definition is purely truth-functional and Frege noted the discrepancy between this truth-functional definition and ordinary uses of the word 'if'. He needed the Philonian conditional to formulate the rule of detachment.

Negation was expressed by Frege by adding a small vertical stroke to the sign of contents: —┬— A. There were no special symbols for other connectives – they were simply expressed as combinations of symbols for negation and implication. Hence for example the equivalence 'A if and only if B' was written symbolically as follows:

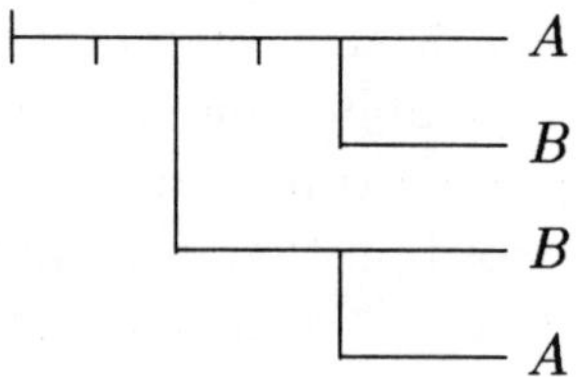

The circumstance that a function Φ is a fact for every value of the argument was expressed in the following way:

$$\overset{\mathfrak{a}}{\smile}\!\!-\!\!-\; \Phi(\mathfrak{a})$$

(note that Frege used here gothic letters as bounded variables). The existential quantifier was expressed as a combination of symbols for negation and the universal quantifier.

Frege introduced also a symbol for identity of content. If Γ and Δ are names of any kind, i.e. not necessarily propositional signs, then the expression $\vdash (\Gamma \equiv \Delta)$ means 'The name Γ and the name Δ have the same conceptual content, so that Γ can always be replaced by Δ and conversely'. Note that the judgement $\vdash (\Gamma \equiv \Delta)$ is about names (signs) and not about their content.

One can formulate strong arguments against such a conception. The recognition of them has forced Frege to split later the notion of conceptual content into sense (*Sinn*) and reference (*Bedeutung*) (cf. Frege 1891, 1892).

6.3.4. Having introduced the symbolism Frege developed his system of logic. It is based on the following nine axioms:

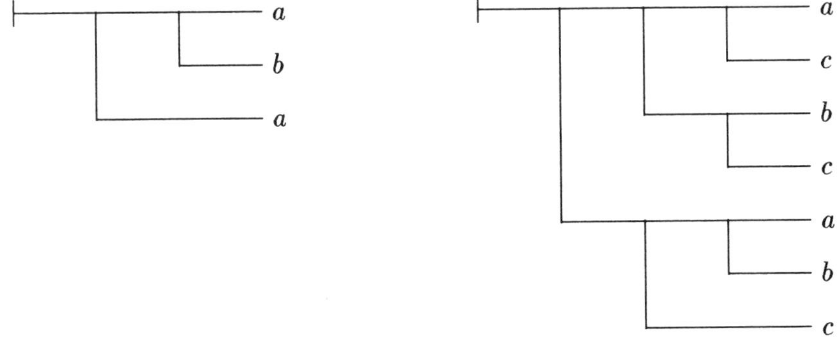

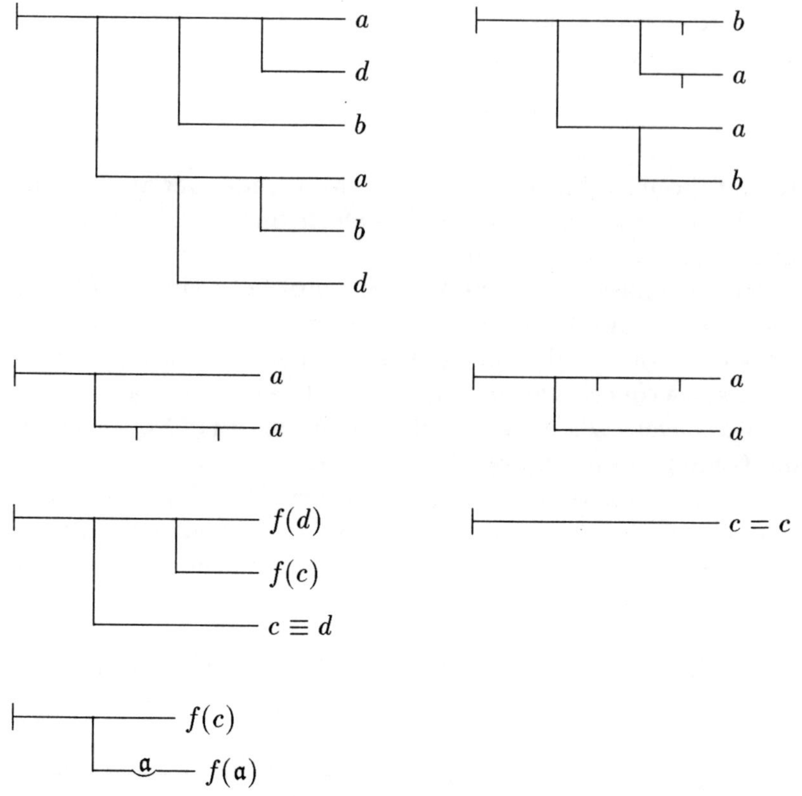

In modern symbolism they would be written as follows:

(1) $p \longrightarrow (q \longrightarrow p)$,
(2) $[p \longrightarrow (q \longrightarrow r)] \longrightarrow [(p \longrightarrow q) \longrightarrow (p \longrightarrow r)]$,
(3) $[p \longrightarrow (q \longrightarrow r)] \longrightarrow [q \longrightarrow (p \longrightarrow r)]$,
(4) $(p \longrightarrow q) \longrightarrow (\neg q \longrightarrow \neg p)$,
(5) $\neg\neg p \longrightarrow p$,
(6) $p \longrightarrow \neg\neg p$,
(7) $c = d \longrightarrow f(c) = f(d)$,
(8) $c = c$,
(9) $\forall x f(x) \longrightarrow f(a)$.

Frege stated that he used only one principle of inference in derivations of theorems from the axioms, but in fact he used four principles. The rule explicitly mentioned by him is the rule of detachment. It explains why he has chosen just negation and conditional as the only primitive connectives – he wanted to preserve the simple formulation of the rule of detachment. Beside the latter Frege used in *Begriffsschrift* also an unstated rule of substitution, the rule of generalization and the rule leading from $A \longrightarrow F(x)$ to $A \longrightarrow \forall x F(x)$ (provided that x does not occur free in A).

In the framework of this system Frege proved various theorems giving detailed proofs of them. Those proofs, even in the absence of some explicit rules, can be unambiguously reconstructed. His axioms together with the four rules form indeed a complete set of axioms in the technical sense of that phrase. Frege himself did not raised, of course, any question of completeness or consistency or independence (cf. 6.4.6 below). Note that the third of Frege's axioms is redundant, since it can be derived from the first two. It is also possible to replace, in the presence of (1) and (2), the axioms (4), (5) and (6) by a single new axiom for negation, namely:

$$(\neg p \longrightarrow \neg q) \longrightarrow (q \longrightarrow p).$$

Frege used the whole machinery presented above to develop (in the third part of *Begriffsschrift*) a theory of mathematical sequences. He introduced the relation called later by Whitehead and Russell the ancestral relation. It served him for the justification of mathematical induction in *Grundlagen der Arithmetik* (cf. Frege 1884) and generally to the logical reconstruction of arithmetic – what was his main purpose in *Grundgesetze der Arithmetik* (cf. Frege 1893 and Frege 1903). In this way Frege became the founder and the first representative of logicism in the philosophy of mathematics.

6.3.5. Though Frege's system constructed in *Begriffsschrift* has turned out to be inconsistent – this was discovered in 1901 by Russell who built in this system a paradox of the class of all classes that do not contain themselves as elements, known today as Russell's paradox – nevertheless Frege's achievements remain really great. *Begriffsschrift* was the first comprehensive system of formal logic and it contains all the essentials of modern logic. Its main novelties were the analysis of the proposition into function and argument(s), rather than subject and predicate, the quantification theory, the truth-functional propositional calculus and, what is the most important for us here, a system of logic in which derivations are carried out exclusively according to the form of the expressions.

Frege repeatedly opposed a *lingua characteristica* to a *calculus ratiocinator*. In works of his predecessors calculus of propositions and classes, logic, translated into formulas, was studied by means of arguments resting upon an intuitive logic. Frege constructed logic as a language that did not need to be supplemented by any intuitive reasoning. Therefore he was very careful to describe his system in purely formal terms (speaking for example of letters rather than of variables to avoid any imprecision and ambiguity).

6.3.6. The reception of Frege's work was rather bad. One of the reasons was certainly his symbolism. Though it has many advantages – it facilitaties inferences by detachment, allows to perceive the structure of a formula at a glance or to perform substitutions with ease – it has also some disadvantages. One of them is the fact that it occupies a great deal of space and is not easy to print. Frege himself said that "the comfort of the typesetter is certainly not the *summum bonum*" (cf. Frege 1896), but it cannot be denied that his symbolism very often hindered to grasp the importance of his ideas or even to understand the contents of his papers (cf. e.g. Russell 1903). On the other hand one should admit that Frege presented too many difficult novelties at once and with too little concessions to human weakness. Logicians have recognized the merits of Frege's ideas only when they themselves became ready to discuss the questions he raised.

The first who gave serious recognition to Frege's works was G. Peano. They corresponded with each other, exchanged copies of their papers, exchanged comments and remarks on them. The real reception of the ideas of Frege began from Russell who learned of Frege and his works through Peano (cf. the prevous section). Russell studied Frege's papers very carefully and appreciated their importance. This had deep consequences and strongly influenced the development of logic – we shall discuss it in the next section.

6.4. B. Russell and the fulfillment of Peano's and Frege's projects

6.4.1. The projects of Peano and Frege presented in the previous sections found their fulfilment in the monumental work *Principia Mathematica* written by two English mathematicians and philosophers: Alfred North Whitehead (1861–1947) and Bertrand Russell

(1872–1970), and published in three volumes in the years 1910–1913. It is a detailed study of logic and set theory and a construction, on that basis, of the classical mathematics.

In June 1901 when studying the first volume of Frege's *Grundgesetze der Arithmetik* Russell discovered the paradox of the class of all classes that do not contain themselves as elements (called today Russell's paradox). This meant that the system of logic to which Frege wanted to reduce mathematics was inconsistent. Russell communicated his discovery to Frege on 16 June 1902 (cf. Russell 1902 and Frege 1902). In this situation the task of the reduction of mathematics to logic should be resumed.

6.4.2. Discussing 'the contradiction', as he called it, in *The Principles of Mathematics* (cf. Russell 1903), Russell mentioned, without much elaboration, that "the class as many is of a different type from the terms of the class" and that "it is the distinction of logical types that is the key to the whole mystery". He examined also other solutions but found them less satisfactory.

Soon after publishing *The Principles of Mathematics* Russell abandoned the theory of types. In a paper written in 1905 (cf. Russell 1907) he proposed three theories to overcome the difficulties raised by the paradoxes (at that time Russell knew of other paradoxes as well, for instance the Burali-Forti paradox and that of the greatest cardinal). They were: (1) the zigzag theory ("propositional functions determine classes when they are fairly simple, and only fail to do so when they are complicated and recondite"), (2) the theory of limitation of size ("there is not such a thing as the class of all entities"), (3) the no-classes theory ("classes and relations are banished together"). Notice that the theory of types is not mentioned here. Add also that Russell believed that the no-classes theory "affords the complete solution to all the difficulties" (cf. a note added on 5 February 1906 to the paper Russell 1907).

6.4.3. Those hopes disappeared soon and Russell turned back to the theory of types and proceeded to develop it in detail. It was done in a paper "Mathematical logic as based on the theory of types" published in 1908 and later in a monumental work *Principia Mathematica* written together with his teacher A. N. Whitehead and published in the years 1910–1913 (vol. I 1910, vol. II 1912, vol. III 1913).

Russell saw the key to the paradoxes in what he called the vicious-circle principle: "No totality can contain members defined in terms of itself". This is connected with the principle of Poincaré forbidding impredicative definitions of objects, which Russell incorporated into his system. To realize it Russell saw the universe as deviding into levels, or types. All conditions and properties we can consider form an infinite hierarchy of types: properties of the first type are properties of individuals, properties of the second type are properties of properties of individuals etc. We can speak of all the things fulfilling a given condition only if they are all of the same type. These intuitions led to the ramified theory of types developed by A. N. Whitehead and B. Russell. It made possible the elimination of paradoxes. This theory, rather complicated and formulated *ad hoc*, was simplified later by F. P. Ramsey (cf. Ramsey 1925) and L. Chwistek (cf. Chwistek 1921, 1922, 1924–25) and is known today as the simple theory of types.

6.4.4. The theory of types was the basis of the system of *Principia Mathematica.* Whitehead and Russell's aim they wanted to realize in it was the reduction of the pure mathematics to logic. They used of course some ideas and achievements of Peano and Frege. The whole work was written in a symbolic language (and the usage of the colloquial language was reduced to the minimum), the symbolism of Peano was adopted rather than that of Frege. The derivations of theorems were performed with the same rigour as it was by Frege. The system of logic of *Principia* was the propositional calculus, adorned with other features such as the theory of definite descriptions, introduced there especially to define mathematical functions. Set theory was widely used, but a set X was reduced to a propositional function *via* the contextual definition of a property (or functional function) which X satisfied. The propositional calculus was based on two primitive notions: negation and disjunction, all other connectives were defined in terms of those two, namely (in the notation of *Principia*):

$$p \supset q. = . \sim p \vee q,$$
$$p.q. = . \sim (\sim p \vee \sim q),$$
$$p \equiv q. = .p \supset q.q \supset p.$$

The basic axioms were:

(A) the propositional calculus:

(1) $(p \vee p) \supset p$,
(2) $q \supset (p \supset q)$,
(3) $(p \vee q) \supset (q \vee p)$,
(4) $[p \vee (q \vee r)] \supset [(p \vee q) \vee r]$,
(5) $(q \supset r) \supset [(p \vee q) \supset (p \vee r)]$,

(B) the quantification theory (x, y are here variables of the same type and x is not free in p):
(1) $(x)Fx \supset Fy$,
(2) $(x)(p \vee Fx) \supset (p \vee (x)Fx)$,

(C) the axiom of reducibility,
(D) identity,
(E) the axiom of infinity,
(F) the axiom of choice.

Łukasiewicz in 1925 and Bernays in 1926 observed that the axiom (A4) could be derived from the others and therefore it is superfluous.

The basic rules of inference were the rule of detachment (i.e. modus ponens), the generalization, and the (unspoken but used) rule of substitution.

Remark that there is a certain inelegance in this system. Since the logical signs of negation and disjunction are to be taken as primitive, they should be used in the formulation of the axioms and rules of inference to the exclusion of all other signs. Of course, one can replace $p \supset q$ by $\sim p \vee q$ in those axioms but they become then rather complicated.

6.4.5. The main merit of Whitehead and Russell's *Principia Mathematica* being the fulfilment of Peano's and Frege's projects was the reduction of the pure mathematics to logic. It was performed in a formal and rigorous way by reducing all reasonings to the rules of inference stated at the very beginning.

Peano and Frege constructed symbolic languages making possible the axiomatization (Peano) and formalization (Frege) of mathematics and mathematical reasonings. Russell and Whitehead improved their systems eliminating difficulties and paradoxes they met, improving their symbolism and carring out the formal development of mathematics.

6.4.6. It is worth noting here that for Russell every logical formula had a fixed meaning, hence there was no question of reinterpreting any sign (as it was the case by the algebraists). For Russell (and for Frege as well) logic was about something, namely everything. The laws of logic have content: they are the most general truths about the logical furniture of the universe. Russell wrote: "Logic is concerned with the real world just as truly as zoology, though with its more abstract and general features" (cf. Russell 1919). Observe also that Frege's and Russell's conception of logic was all-embracing in that it included indiscriminately both logical and metalogical topics. Therefore neither Frege nor Russell raised any serious metasystematic questions at all. For them logic was the system — the results of logic were simply the logical truths, and were to be arrived at by deriving them in the system.

6.5. Skolemization

6.5.1. To present the idea of skolemization which formed another 20th century's step on the way to mechanization of reasonings we must say some words about a German mathematician and logician Leopold Löwenheim (1878–1957) and his paper from 1915. We mean here the paper "Über Möglichkeiten im Relativkalkül" which dealt with a problem connected with the validity, in different domains, of formulas of what we call today the first order predicate calculus and with various aspects of the reduction and the decision problems. Observe that all these topics had remained alien to the trend in logic created by Peano, Frege and Russell – the trend becoming then dominant. Nevertheless the Löwenheim's paper may be rightly considered as a pioneer work in logic because the problems he had been considering came soon into the foreground.

Löwenheim was squarely in Schröder's school, i.e. in the algebraic tradition. He used in the paper we are discussing here Schröder's notation, he hold that all of mathematics can be framed within the relative calculus. We read in Löwenheim (1915): "Every theorem of mathematics ...can be written as a relative equation; the mathematical theorem then stands or falls according as the equation is satisfied or not. This transformation of arbitrary mathematical theorems into relative equations can be carried out, I believe, by anyone who knows the work of Whitehead and Russell". In a later paper (cf. Löwenheim 1940) he argued that mathematics can be "Schröderized". We can suspect that he did not fully grasp the notion that by means of formal representation we may fully exhibit the intuitive content of mathematical discourse and the inferential relations among mathematical sentences.

6.5.2. Theorem 2 in §2 of Löwenheim's paper from 1915 is the now famous Löwenheim's theorem. It states that if a finitely valid well-formed formula is not valid, then it is not $\aleph_0$-valid. In other words if a well-formed formula is $\aleph_0$-valid then it is valid provided there is not some finite domain in which it is invalid.

To prove this theorem Löwenhheim shows first of all that for any given first-order formula one can obtain a formula that has a certain normal form and is equivalent, so far as satisfiability is concerned, to the given formula. Observe that he shifts here from validity to satisfiability. The reason is the fact that the formula consitutes the left side of an equation whose right side is 0 and to write that a formula is equal to 0 amounts to writing a negation sign in front of the formula. The formula in normal form contains existential quantifiers, universal quantifiers and a quantifier-free expression. The number of existential quantifiers depends on the number of individuals in the *Denkbereich,* in particular for infinite domains there are infinitely many quantifiers and we get an infinitely long expression.

Note here also that §3 of Löwenheim (1915) contained the positive solution of the decision problem for the singulary predicate calculus of first order with identity. The proof of the result was later simplified by Skolem and Behmann (cf. Skolem 1919 and Behmann 1922). §4 presented the reduction of the decision problem for the first-order predicate calculus to that of its binary fragment. This result was later sharpened by Herbrand and Kalmár (cf. Herbrand 1931, Kalmár 1932, 1934, 1936).

6.5.3. In 1920 the Löwenheim's theorem was extended by the Norwegian mathematician and logician Thoralf Skolem (1887–1963). Skolem generalized it to denumerably infinite sets of formulas and proved that if a countable set F of first-order propositions is satisfiable, i.e. has a model, then there exists a denumerable submodel of this model in which all formulas of F are satisfied.

6.5.4. To demonstrate the theorem Skolem proved that every formula of the first-order predicate calculus has a special normal form that is called today a Skolem normal form for satisfiability. Such a form consists of a string of universal quantifiers, a string of existential quantifiers and then a quantifier-free part. It was shown that a formula is satisfiable in a given domain if and only if its Skolem normal form for satisfiability is satisfiable in the domain. The denumerable submodel needed in the theorem was obtained by Skolem from the original model by a process of thinning out. This process was performed with the help of the axiom of choice. The usage of the normal form of a formula enabled Skolem to dispense with the infinite strings of quantifiers considered by Löwenheim. Note that in 1929 Skolem gave a simplified version of the proof (using again the axiom of choice, more exactly König's lemma which can be regarded as a version of the axiom of dependent choices – cf. Skolem 1929).

Skolem normal forms of formulas have become one of the logician's standard tools. They constitute a reduction class for quantification theory.

In 1922 T. Skolem published a paper "Einige Bemerkungen zur axiomatischen Begründung der Mengenlehre" in which he proved a weaker version of his theorem not using this time the axiom of choice. He showed that if a set F of formulas is satisfiable then it is satisfiable in a denumarable universe (i.e. in the domain of natural numbers). There was no mention of submodels. To prove this Skolem first set up effective functions on the integers and then showed, by considering numerical quantifier-free instances of the given quantified formula, that these functions can act as evaluations for the existential variables.

6.5.5. From our point of view (i.e. from the point of view of the development of the idea of formalizing and mechanizing the reasonings) the most important is Skolem's paper from 1928. It was a lecture delivered before the Norwegian Mathematical Society (on 22nd October 1928). Skolem proposed there a proof procedure for the first-order predicate calculus. To describe it consider an arbitrary closed formula φ in the prenex form (i.e. φ consists of a string of universal and existential quantifiers and then of a quantifier-free matrix). We put it into what one calls today its functional form for satisfiability – all quantifiers are dropped and the occurrences of each variable bound by an existential quantifier are replaced by a functional term containing the variables that in φ are bound by universal quantifiers preceding the existential quantifier. For example, if φ is of the form $(x)(y)(Ez)(u)(Ew)\psi(x,y,z,u,w)$, where ψ is quantifier-free then the functional form of φ, denoted by φ', will be:

$$\psi(x,y,f(x,y),u,g(x,y,u)),$$

where $f(x,y)$ and $g(x,y,u)$ are functional terms (to be honest we should note that Skolem did not use this notion), f and g being new function symbols.

Introduce now constants of various levels. 0 is the constant of the 0th level. Constants of the nth level are the expressions obtained when in each functional term of φ', the functional form of φ, the variables are replaced by constants that include at least one constant of, but no constant beyond, the $(n-1)^{\text{th}}$ level. In this way we can generate from 0 the lexicon of φ' (the term 'lexicon' was introduced by Quine, cf. Quine 1955). The subset of the lexicon of φ consisting of all consta nts of levels $\leq n$ we shall call the lexicon of level n.

Take now a formula χ which is the functional form φ' of φ or any truth-functional part of φ'. By a lexical instance of χ we shall mean the formula obtained from χ by replacing all the variables of χ by elements from the lexicon of φ'. A solution of the nth level is an assignment of truth-values to lexical instances of atomic parts of φ' such that it verifies the conjunction of all the lexical instances of φ' formed from the lexicon of $(n-1)^{\text{th}}$ level of φ'.

It can be seen that: 1° any solution of the nth level is an extension of a solution of the $(n-1)^{\text{th}}$ level, 2° either (*i*) for some n there is no solution of the nth level or (*ii*) for every n there is a solution of the nth level. In the case (*i*) φ is not satisfiable (Skolem used here the phrase 'contains a contradiction' – 'enthält einen Widerspruch'). In this way we get that if φ is satisfiable then for every n there is a solution of the nth level. In papers from 1922 and 1929 Skolem obtained also the converse, namely if for every n there is a solution of the nth level, then φ is satisfiable (in fact he actually proved that φ is $\aleph_0$-satisfiable). In this way one gets a proof procedure for the first-order predicate calculus: given a formula φ in prenex form and a number n, we can effectively decide whether or not there is a solution of the nth level for φ; if for some n there is no solution of the nth level, then φ is not satisfiable, hence $\neg\varphi$ is valid. Observe that this approach is an alternative to the axiomatic one developed by Peano, Frege and Russell (cf. sections 6.2, 6.3 and 6.4) and continued by Hilbert and his school (see below). Note also that Skolem was a constant oponent of all formalist and logicist foundational programs. He insisted that the foundations of mathematics lay in what he called 'the recursive mode of thought' (cf. Skolem 1923).

Add that in the paper from 1928 Skolem considered also the decision problem from the point of view of his proof procedure. Namely for certain classes of formulas one can effectively decide whether or not for every n there is a solution of the nth level. An example of such a class is the class of formulas whose prefix consists of a single universal quantifier followed by any number of existential quantifiers (a subclass of this class consisting of formulas of the form $(x)(Ey)\chi$, χ being quantifier-free was considered earlier by Bernays and Schönfinkel (1928), the whole class was studied independently and simultaneously with Skolem by Ackermann whose results go beyond that of Skolem, cf. Ackermann 1928). Skolem considered also the subclass of formulas of the form $(x_1)\dots(x_m)(Ey_1)\dots(Ey_n)\chi$ where χ is quantifier-free and such that all variables $x_1,\dots,x_m$ occur in each of the atomic parts of it.

6.5.6. To sum up we may say that Skolem was interested in analyzing statements in terms of truth-functional logic. Towards this end he introduced a notion of expansion. Being inspired by Löwenheim (1915) he eliminated the need for first fixing a universe over which to expand a formula and considered expansions which were finitely long (whereas Löwenheim's were not, in fact they could even be "uncountably long"). An expansion of a formula was simply a finite conjunction of quantifier-free instances of the formula obtained by following certain instantiation rules which reflect the idea of making successive choices (in the sense of picking symbols).

6.6. D. Hilbert and his program

6.6.1. As we have mentioned above Skolem saw the foundations of mathematics in "the recursive mode of thought". In a opposite direction were working the German mathematician David Hilbert (1862–1943) and his school. Hilbert's interest in foundations dates back to his investigations of geometry – cf. Hilbert (1899). In his famous lecture at the International Congress of Mathematicians in Paris in 1900 he mentioned the consistency of analysis as a second problem on his list of 23 main problems in mathematics that should be solved (cf. Hilbert 1900). Hilbert elaborated on this problem later but in a rather obscure manner – cf. Hilbert (1904). This paper is important also for other reasons: namely Hilbert introduced here the themes that he was going to develop, modify or make more precise in his further works.

He returned to foundational questions in 1918. At that time a great advance had been made in formal methods. Particularly influential was *Principia Mathematica* which Hilbert called "the crowning achievement of the work of axiomatization". But Hilbert's view of this achievement was quite different from Russell's. As we pointed out earlier Russell claimed that *Principia* contained a system of logic on which all mathematical reasonings should (and can) be based. It had an epistemological priority. For Hilbert *Principia* provided an empirical evidence of the possibility of formalization. This work showed that all mathematical results can be captured by appropriate formal systems. Hilbert claimed that we can think of the content of much of mathematics as given by its formal representation.

6.6.2. Hilbert used this possibility in his famous program of proving the consistency of mathematics, called today Hilbert's program. His proposal was Kantian in character. He wrote in (1926) as follows:

> Kant taught – and it is an integral part of his doctrine – that mathematics treats a subject matter which is given independently of logic. Mathematics, therefore, can never be grounded solely on logic. Consequently, Frege's and Dedekind's attempts to so ground it were doomed to failure.
>
> As a further precondition for using logical deduction and carring out logical operations, something must be given in conception, viz., certain extralogical concrete objects which are intuited as directly experienced prior to all thinking. For logical deduction to be certain, we must be able to see every aspect of these objects, and their properties, differences, sequences, and contiguities must be given, together with the

> objects themselves, as something which cannot be reduced to something else and which requires no reduction. This is the basic philosophy which I find necessary not just for mathematics, but for all scientific thinking, understanding and communicating. The subject matter of mathematics is, in accordance with this theory, the concrete symbols themselves whose structure is immediately clear and recognizable.

Such concrete objects are just natural numbers considered as numerals (certain systems of symbols): 1, 11, 111, ... One can exactly describe them and relations between them. The part of mathematics talking about those objects is certainly consistent (because facts cannot contradict themselves). But in mathematics, beside such finitistic, real theorems describing concrete objects, we have also infinitistic, ideal ones talking about the actual infinity (to which no real objects correspond). And therefore mathematics needs a justification and foundations. The convincing proof of the consistency of mathematical theory ought to be a finitistic one (i.e. a proof using no ideal assumptions). Hilbert thought that such a proof was possible and proposed a program of providing it. It consisted of two steps. The first step was just the formalization of mathematics (Hilbert, first of all, thought here about arithmetic, analysis and set theory). It ought to be carried out by fixing an artificial symbolic language and rules of building in it well-formed formulas. Further axioms and rules of inference ought to be fixed (the rules could refer only to the form, to the shape of formulas and not to their sense or meaning). In such a way theorems of mathematics become those formulas of our formal language which have a formal proof based on a given set of axioms and given rules of inference. There was one condition put on the set of axioms: they ought to be chosen in such a way that they suffice to solve any problem formulated in the language of a considered theory, i.e. they ought to form a complete set of axioms. The second step in Hilbert's program was now to give a proof of the consistency of mathematics. Such a proof could be carried out by finitistic methods because it was enough to consider formal proofs, i.e. sequences of symbols, and to verify if there were two sequences such that one of them finishes with formula φ and the other with formula $\neg\varphi$. If there were such proofs then mathematics would be inconsistent, if not, then it would be consistent. But the study of formal proofs deals with finite, concrete objects (namely sequences of symbols formed according to some rules) and hence is finitistic.

Observe that for Hilbert formalization was just the way in which we, as finite intelligences, can in fact deal with the infinite. We attain to the infinite by means of signs. He recognized in quantifiers that constituent of formal systems that allows us to do this.

6.6.3. Hilbert has been working on his project for several years. The most comprehensive publication in this direction is the two volume monograph published with P. Bernays *Grundlagen der Mathematik* (1934, 1939) which explained the situation and set forth the results achieved up to the time of publication.

Hilbert based his system of logic on the following set of axioms – cf. Hilbert (1928), Hilbert-Bernays (1934, 1939) (we write the axioms in the original Hilbert's notation):

I. axioms of implication:

1. $A \to (B \to A)$,
2. $(A \to (A \to B)) \to (A \to B)$,
3. $(A \to B) \to ((B \to C) \to (A \to C))$,

II. axioms of conjunction:

4. $A \& B \to A$,
5. $A \& B \to B$,
6. $(A \to B) \to ((A \to C) \to (A \to B \& C))$,

III. axioms of disjunction:

7. $A \to A \vee B$,
8. $B \to A \vee B$,
9. $(A \to C) \to ((B \to C) \to (A \vee B \to C))$,

IV. axioms of equivalence:

10. $(A \sim B) \to (A \to B)$,
11. $(A \sim B) \to (B \to A)$,
12. $(A \to B) \to ((B \to A) \to (A \sim B))$,

V. axioms of negation:

13. $(A \to B) \to (\neg B \to \neg A)$,

14. $A \to \neg\neg A$,

15. $\neg\neg A \to A$,

VI. axioms for quantifiers:

16. $\forall x A(x) \to A(a)$,

17. $A(a) \to \exists x A(x)$.

The following three rules of inference were allowed:

(i) the rule of detachment,

(ii) $\dfrac{\mathcal{A} \to \mathcal{B}(a)}{\mathcal{A} \to \forall x \mathcal{B}(x)}$

(iii) $\dfrac{\mathcal{B}(a) \to \mathcal{A}}{\exists x \mathcal{B}(x) \to \mathcal{A}}$

where x does not occur in $\mathcal{B}(a)$ and the variable a does not occur in $\mathcal{A}$.

In fact Hilbert needed also the principle of substitution in derivations of theorems from axioms. Note that the given set of axioms is really perspicuous and natural.

6.6.4. In the early days of his program Hilbert introduced also a simplifying device in the formal systems, called today Hilbert-terms (cf. Hilbert 1928). He regarded them as auxiliary, proof-theoretical notations for use in obtaining proofs and shortening their presentation.

A term $\varepsilon_x A(x)$ was to denote an element x of which A holds, provided there are some. If A contains free variables then $\varepsilon_x A(x)$ represents a function of those variables which picks out a suitable value for x given any values for those variables. In this way the very notation exhibits the choice-like nature of quantification. Hilbert stated the following axiom:

$$A(a) \to A(\varepsilon_x A(x))$$

and indicated that the quantifiers 'for all' and 'there exists' can be defined by means of ε, namely $\exists x A(x)$ as $A(\varepsilon_x A(x))$ and $\forall x A(x)$ as $A(\varepsilon_x(\neg A(x))$.

Observe that this approach resembles the Skolem's approach exploiting the similarities between the quantifier (or, more exactly, quantifier-dependence) and choice functions, but they went in different directions: Skolem used these similarities to denigrate the power of quantification and Hilbert was impressed with how much the quantifier does, insofar as it acts like a choice function.

Hilbert indicated a purely auxiliary character of Hilbert-terms showing that if an 'ordinary' sequent is proved with Hilbert-terms occurring in the proof, it can be proved without them occurring in the proof.

6.6.5. Though the program of Hilbert could not be fully realized (the incompleteness results of K. Gödel from 1931 showed that it is impossible), nevertheless the study started by Hilbert has been extremely fruitful. As W. and M. Kneale write: "Apart from the problem of consistency, there are many questions about deductive systems for which the technique [Hilbert] suggested seems appropriate, in particular questions about the completeness of axiom systems and about the possibility of devising decision procedures, or rules of thumb, for the solution of problems in various branches of mathematics. In all such investigations it is essential that the theory under investigation should be formalized by the help of a strict logical symbolism, and so it may be said that here at last the work of Leibniz and Frege has produced results of importance to the progress of mathematics, though not perhaps the results they expected" (Kneale 1962, p. 686).

In the next sections we shall indicate the influence of Hilbert's program on various further investigations, in particular on the work of Herbrand and Gentzen.

6.7. J. Herbrand

6.7.1. One of the most important results of investigations generated by Hilbert's program is the fundamental theorem of Jacques Herbrand (1908–1931). It is contained in Chapter 4 of his doctoral thesis (finished in April 1929, presented to the Sorbonne – the defence took place on 11 June 1930, and published in 1930).

The work of Herbrand can be viewed as a reinterpretation, from the point of view of Hilbert's program, of the results of Löwenheim and Skolem (cf. section 6.5). Herbrand wrote about his fundamental theorem: "[it is] a more precise statement of the well-known Löwenheim-Skolem theorem" (cf. Herbrand 1931). Having adopted Hilbert's finitistic viewpoint, Herbrand spoke not about satisfiability and validity, as Löwenheim and Skolem did, but used syntactic notions such as provability and (in)consistency.

6.7.2. The fundamental theorem contains a reduction (in a certain sense) of predicate logic to propositional logic, more exactly it shows that a formula is derivable in the axiomatic system of quantification logic if and only if its negation has a truth-functionally inconsistent expansion. Herbrand intended to prove this theorem by finitistic means.

Herbrand considered an axiomatic system of quantification logic similar to the system of *Principia Mathematica* of Whitehead and Russell (1910–1913) and of Hilbert and Ackermann (1928). He proved that it is equivalent to the quantification theory used by Whitehead and Russell (cf. section 7 of chapter 2 of Herbrand 1930).

It would be unreasonable to give, in a historical work like the present one, a detailed formulation of Herbrand's fundamental result. Hence we shall limit ourselves to a sketchy description of it indicating the main ideas.

Herbrand correlates with each formula Z of the predicate calculus (he used in fact letters like p as syntactic variables ranging over propositions but we find it more convenient to use here capital letters) an infinite sequence of propositional formulas having the form of a disjunction:

$$H_n(Z) = H_n = A_1 \vee \ldots \vee A_n.$$

These formulas are called today the Herbrand disjunctions. The definition of H_n is effective, i.e. H_n can be obtained from Z and n by means of a fixed algorithm. One can establish now the following relationship: Z is provable in the predicate logic if and only if there is an n such that H_n is a theorem of the propositional calculus.

To give the idea of how $H_n(Z)$ is constructed consider the following example. Assume that Z is $(x)(Ey)(z)M(x,y,z)$ where M is quantifier free. Let φ be an arithmetical function such that $\varphi(i) > i$ and $\varphi(i) < \varphi(j)$ for all integers i, j such that $1 \leq i < j$. Form the disjunction

$$(*) \qquad \bigvee_{i \leq n} M(x_1, x_i, x_{\varphi(i)})$$

and replace in it every atomic formula by a propositional variable in such a way that different atomic formulas are replaced by different propositional variables. The resulting formula is the nth Herbrand disjunction.

One may interpret the formula in the following way (cf. Kreisel 1953-54): suppose that we are trying to build a counterexample to the formula Z. We will then look for an element a and a function f correlating with each p an element $f(p)$ such that $\neg M(a, p, f(p))$ is true. Let us substitute for p arbitrary values $p_1, \ldots\ p_n$. Hence the conjunction:

$$\bigwedge_{i \leq n} \neg M(a, p_i, f(p_i))$$

is true. The formula $(*)$ may thus be interpreted as a statement that whatever our choice of a and of the function f will be, our attempt to build a counterexample to Z will fail in the field of at most n elements.

We should prove now that $(*)$ is provable in the propositional calculus if and only if Z is provable in the predicate logic. One part of this equivalence, namely the implication from the left to the right is rather easy to prove. The proof of the other part is more difficult. Note that proving it Herbrand came very close to the discovery of the completeness theorem. One can suspect that he did not make the decisive step only because he was constrained by the finitistic attitude of Hilbert's program.

6.7.3. Remark that Herbrand's proof of the fundamental theorem was fallacious. In 1939 Bernays wrote that "Herbrand's proof is hard to follow" (cf. Hilbert-Bernays 1939, footnote 1, p.158). Gödel claimed in 1963 that in the early forties he had discovered an essential gap in Herbrand's argument but he never published anything on this subject. A detailed analysis of errors in Herbrand's paper was published by Dreben, Andrews and Aanderaa in 1963 (cf. Dreben, Andrews, Aanderaa 1963 and 1963a, Dreben 1963, Dreben, Aanderaa 1964; cf. also Dreben, Denton 1966). Note that a correct, strictly finitistic proof of the fundamental theorem was given by Bernays (cf. Hilbert-Bernays 1939).

6.7.4. A detailed analysis of the proof of Herbrand's theorem shows that we are able to get rid of the rule of modus ponens altogether. Herbrand considered this fact as an important one. It implies that if a formula Z is provable in the predicate calculus then there exists a proof of it consisting exclusively of subformulas of Z. This is connected with the *Hauptsatz* of G.Gentzen who rediscovered and greatly improved many of Herbrand's results (we shall discuss his achievements in the next section).

6.7.5. Herbrand showed an example of applications of his fundamental theorem in the paper "Sur la non-contradiction de l'arithmétique" published in 1931 (cf. Herbrand 1931a, the manuscript of it was received by the editors on 27 July 1931 – on the same day Herbrand had an accident while climbing in the Alps and was killed in a fall). He proved there consistency of a fragment of arithmetic in which the formula that can be substituted into the induction scheme does not contain any quantifiers or, if it does, does not contain any function but the successor function. This result as well as the fundamental theorem discussed above, i.e. the analysis of quantification in terms of truth-functional properties of expansions can be treated without any doubt as an important contribution to the realization of the Hilbert's school program.

6.8. G. Gentzen and natural deduction

6.8.1. Another example of investigations which grew out from Hilbert program is the natural deduction of Gerhard Gentzen (1909–1945). He started his scientific career by publishing in 1932 the paper "Über die Existenz unabhängiger Axiomensysteme zu unendlichen Satzsystemen". This relatively unknown paper which has received less attention since its publication than it deserves (Gentzen has shown in it that not all sentence systems possess independent axiom systems, whereas all 'linear' sentence systems do possess an independent axiomatization) reveals another source of inspiration for Gentzen's investigations. It was the work of Paul Hertz. Gentzen used not only Hertz's methodology but also some of his terminology (cf. Gentzen 1935 and later works where we find, for example, the terms 'antecedent' and 'succedent' of a sequent borrowed from Hertz, also the notion 'logical consequence' as used by Gentzen was largely inspired by Hertz's work).

The most important from our point of view is Gentzen's paper "Untersuchungen über das logische Schliessen" from 1935. There he broke away from the traditional formulations of predicate logic as they were developed by Frege, Russell and Hilbert and presented two basically different versions of predicate logic (both in two variants – classical and intuitionistic). It should be noted here that similar attempt was made independently and simultaneously by S. Jaśkowski (1906–1965) (cf. his paper 1934). Jaśkowski published a different version of 'natural deduction' based on ideas first put forward in Łukasiewicz's seminar in 1926.

Gentzen's idea of a 'natural deduction' came from detailed analysis of Euclid's classical proof of the nonexistence of a largest prime number. It revealed to him what properties a system of natural deduction should have and led him to a calculus based on assumption formulae instead of the usual axiom formulae and to the separation of the rules for the logical operators into 'introduction' rules and into 'elimination' rules.

As we have mentioned above, Gentzen (1935) introduced two versions of predicate logic: one, called by him natural deduction (*der Kalkül des natürlichen Schliessens*) and denoted by NK (classical natural deduction) and NJ (intuitionistic natural deduction) and second called logistic calculi (again in two variants: the classical LK and intuitionistic LJ). The second one was called later the sequent calculus or Gentzen's calculus.

6.8.2. We shall first describe the systems NK and NJ. Note that Gentzen used in these systems two symbols for inference: the usual horizontal line which separated premisses and conclusion and the symbol used in the situation when among premisses there was certain scheme of inference (this is one of characteristic features of the systems of Gentzen). In the latter case we shall use the symbol $A \vdash B$ (Gentzen in 1935 wrote it as $\genfrac{}{}{0pt}{}{[A]}{B}$). Small letters like x, y, z

will denote bounded variables and letters like a, b, c – parameters (free variables), the symbol $\bigwedge$ will denote 'the false proposition'. Gentzen introduced the following inference figure schemata – they were of two types: introduction rules and elimination rules:

	introduction rules		elimination rules
& – I	$\dfrac{A \quad B}{A\&B}$	& – E	$\dfrac{A\&B}{A} \quad \dfrac{A\&B}{B}$
∨ – I	$\dfrac{A}{A \vee B} \quad \dfrac{B}{A \vee B}$	∨ – E	$\dfrac{A \vee B \quad A \vdash C \quad B \vdash C}{C}$
→ – I	$\dfrac{A \vdash B}{A \rightarrow B}$	→ – E	$\dfrac{A \quad A \rightarrow B}{B}$
¬ – I	$\dfrac{A \vdash \bigwedge}{\neg A}$	¬ – E	$\dfrac{A \ \neg A}{\bigwedge} \quad \dfrac{\neg\neg A}{A}$
∀ – I	$\dfrac{A(a)}{\forall x A(x)}$	∀ – E	$\dfrac{\forall x A(x)}{A(a)}$
∃ – I	$\dfrac{A(a)}{\exists x A(x)}$	∃ – E	$\dfrac{\exists x A(x) \quad A(a) \vdash B}{B}$

$$\frac{\bigwedge}{A}$$

In the case of ∀–I and ∃–E we have some restriction on the free object variable a (called the *eigenvariable*): it must not occur in the formula $\forall x A(x)$ nor in any assumption formula upon which that formula depends, and it must not occur in the formula $\exists x A(x)$ nor in B nor in any assumption formula upon which that formula depends, with the exception of the assumption formula $A(a)$.

Those rules constitute the system NK. To obtain the system NJ we eliminate the second rule of $\neg - E$, i.e. the rule $\frac{\neg\neg A}{A}$.

To illustrate how this system of natural deduction works let us give here three examples taken from Gentzen (1935). First we write a proof of the formula $\exists x \forall y A(x,y) \rightarrow \forall y \exists x A(x,y)$:

$$\dfrac{\overset{2}{\exists x \forall y A(x,y)} \qquad \dfrac{\dfrac{\dfrac{\overset{1}{\forall y A(a,y)}}{A(a,b)}\ \forall - E}{\exists x A(x,b)}\ \exists - I}{\forall y \exists x A(x,y)}\ \forall - E}{\dfrac{\forall y \exists x A(x,y)}{\exists x \forall y A(x,y) \to \forall y \exists x A(x,y)}\ \to - I_2}\ \exists - E_1$$

The second example is the proof of the formula:

$$\neg \exists x A(x) \to \forall y \neg A(y).$$

$$\dfrac{\dfrac{\dfrac{\dfrac{\dfrac{\overset{2}{A(a)}}{\exists x A(x)}\ \exists - I \qquad \overset{1}{\neg \exists x A(x)}}{\bigwedge}\ \neg - E}{\neg A(a)}\ \neg - I_2}{\forall \neg A(y)}\ \forall - I}{\neg \exists x A(x) \to \forall y \neg A(y)}\ \to - I_1$$

The last example concerns the formula $(A \vee (B\&C)) \to (A \vee B)\&(A \vee C)$. We proceed as follows:

$$\dfrac{\dfrac{\overset{2}{A \vee (B\&C)} \qquad \dfrac{\dfrac{\overset{1}{A}}{A \vee B}\text{∨-I} \quad \dfrac{\overset{1}{A}}{A \vee C}\text{∨-I}}{(A \vee B)\&(A \vee C)}\ \text{&-I} \qquad \dfrac{\dfrac{\dfrac{\overset{1}{B\&C}}{B}\ \text{&-E}}{A \vee B}\ \text{∨-I} \quad \dfrac{\dfrac{\overset{1}{B\&C}}{B}\ \text{&-E}}{A \vee C}\ \text{∨-I}}{(A \vee B)\&(A \vee C)}\ \text{&-I}}{(A \vee B)\&(A \vee C)}\ \text{-∨E}_1}{A \vee (B\&C) \to (A \vee B)\&(A \vee C)}\ \to -I_2$$

6.8.3. Gentzen wrote in (1935):

> A closer investigation of the specific properties of the natural calculus finally led me to a very general theorem which will be referred to below as the *Hautptsatz*.
>
> The *Hauptsatz* says that every purely logical proof can be reduced to a define, though not unique, normal form. Perhaps we may express the essential properties of such a normal proof by saying: it is not roundabout. No concepts enter into the proof other than those contained in its final result, and their use was therefore essential to the achievement of that result.

In order to formulate precisely the *Hauptsatz* and to prove it Gentzen had to introduce a logical calculus suitable to this purpose, since the calculi NK and NJ proved unsuitable. Therefore he introduced the calculi of sequents LK and LJ. The characteristic feature of this calculus is just the usage of sequents (*Sequenz*). They may be seen as the most general form of a scheme of inference. A sequent is an expression of the form

$$A_1, \ldots, A_n \Rightarrow B_1, \ldots, B_m$$

where $A_1, \ldots, A_n$, $B_1, \ldots, B_m$ are formulas (the symbols $\Rightarrow$ and, are auxiliary and not logical ones). The formulas $A_1, \ldots, A_n$ form the antecedent *(Antezedens)* and the formulas $B_1, \ldots, B_m$ the succedent *(Sukzedens)* of the sequent. Both expressions may be empty. The sequent $A_1, \ldots, A_n \Rightarrow B_1, \ldots, B_m$ can be understood as

$$A_1 \& \ldots \& A_n \Rightarrow B_1 \vee \ldots \vee B_m.$$

If the antecedent is empty, the sequent reduces to the formula $B_1 \vee \ldots \vee B_m$, if the succedent is empty, the sequent means the same as the formula $\neg(A_1 \& \ldots \& A_n)$ or $(A_1 \& \ldots \& A_n) \rightarrow \bigwedge$.

A derivation in the system LK or LJ consists of sequents arranged in a tree form such that it begins always from an initial sequent $A \Rightarrow A$, where A is any formula and in which sequents are transformed one into another according to the following rules which are of two types: structural and operational. To describe them we shall use letters T, U, X, Y, Z to denote arbitrary sequences of formulas separated by commas and letters like A, B, C, D etc. to denote arbitrary formulas. We adopt the following structural rules:

thinning $$\frac{T \Longrightarrow U}{A, T \Longrightarrow U} \qquad \frac{T \Longrightarrow U}{T \Longrightarrow U, A}$$

contraction $$\frac{A, A, T \Longrightarrow U}{A, T \Longrightarrow U} \qquad \frac{T \Longrightarrow U, A, A}{T \Longrightarrow U, A}$$

interchange $$\frac{T, A, B, U \Longrightarrow X}{T, B, A, U \Longrightarrow X} \qquad \frac{T \Longrightarrow U, A, B, X}{T \Longrightarrow U, B, A, X}$$

cut $$\frac{T \Longrightarrow U, A \quad A, X \Longrightarrow Y}{T, X \Longrightarrow U, Y}$$

The operational rules are as follows:

$\& - IS \qquad \dfrac{T \Longrightarrow U, A \quad T \Longrightarrow U, B}{T \Longrightarrow U, A \& B}$

$\& - IA \qquad \dfrac{A, T \Longrightarrow U}{A \& B, T \Longrightarrow U} \qquad \dfrac{B, T \Longrightarrow U}{A \& B, T \Longrightarrow U}$

$\vee - IA \qquad \dfrac{A, T \Longrightarrow U \quad B, T \Longrightarrow U}{A \vee B, T \Longrightarrow U}$

$\vee - IS \qquad \dfrac{T \Longrightarrow U, A}{T \Longrightarrow U, A \vee B} \qquad \dfrac{T \Longrightarrow U, B}{T \Longrightarrow U, A \vee B}$

$\neg - IS \qquad \dfrac{A, T \Longrightarrow U}{T \Longrightarrow U, \neg A}$

$\neg - IA \qquad \dfrac{T \Longrightarrow U, A}{\neg A, T \Longrightarrow U}$

$\rightarrow - IS \qquad \dfrac{A, T \Longrightarrow U, B}{T \Longrightarrow U, A \rightarrow B}$

$\rightarrow - IA \qquad \dfrac{T \Longrightarrow U, A \quad B, X \Longrightarrow Y}{A \rightarrow B, T, X \Longrightarrow U, Y}$

$\forall - IS \qquad \dfrac{T \Longrightarrow U, A(a)}{T \Longrightarrow U, \forall x A(x)}$

$\exists - IA \qquad \dfrac{A(a), T \Longrightarrow U}{\exists x A(x), T \Longrightarrow U}$

$\forall - IA \qquad \dfrac{A(a), T \Longrightarrow U}{\forall x A(x), T \Longrightarrow U}$

$\exists - IS \qquad \dfrac{T \Longrightarrow U, A(a)}{T \Longrightarrow U, \exists x A(x)}$

There is a restriction on the *eigenvariable* a in the schemata $\forall - IS$ and $\exists - IA$: it must not occur in the lower sequent of the inference figure.

The rules just described constitute the classical logistic calculus LK. To obtain the intuitionistic logistic calculus LJ we add one restriction – in the succedent of any sequent no more than one formula may occur.

Following Gentzen (1935) we give two examples of derivations in LK and LJ, respectively:

(1) a derivation in LK – it is a derivation of the law of the excluded middle:

$$
\dfrac{\dfrac{\dfrac{\dfrac{\dfrac{A \Longrightarrow A}{\Longrightarrow A, \neg A}\ \neg - IS}{\Longrightarrow A, A \vee \neg A}\ \vee - IS}{\Longrightarrow A \vee \neg A, A}\ \text{interchange}}{\Longrightarrow A \vee \neg A, A \vee \neg A}\ \vee - IS}{\Longrightarrow A \vee \neg A}\ \text{contraction}
$$

(2) a derivation in LJ:

$$
\dfrac{\dfrac{\dfrac{\dfrac{\dfrac{A(a) \Longrightarrow A(a)}{A(a) \Longrightarrow \exists x A(x)}\ \exists\text{-}IS \qquad \dfrac{\dfrac{\exists x A(x) \Longrightarrow \exists x A(x)}{\neg \exists x A(x), \exists x A(x) \Longrightarrow}\ \neg\text{-}IA}{\exists x A(x), \neg \exists x A(x) \Longrightarrow}\ \text{interchange}}{A(a), \neg \exists x A(x) \Longrightarrow}\ \text{cut}}{\neg \exists x A(x) \Longrightarrow \neg A(a)}\ \neg\text{-}IS}{\neg \exists x A(x) \Longrightarrow \forall y \neg A(y)}\ \forall\text{-}IS}{\Longrightarrow (\neg \exists x A(x)) \rightarrow (\forall y \neg A(y))}\ \rightarrow\text{-}IS
$$

Gentzen's new approach to the predicate logic turned out to be equivalent to the axiomatic one – in the paper from 1935 Gentzen proved the equivalence of the calculi NJ, NK and LJ, LK with calculi modelled on the formalism of Hilbert, more exactly NJ and LJ are equivalent with the intuitionistic calculus and NK and LK with the classical one.

6.8.4. The calculi LK and LJ have the important property mentioned already above under the name *Hauptsatz.* Gentzen proved namely that the cut-rule can be eliminated from every purely logical proof. Consequently, we obtain the subformula property which says that in a cut-free proof all formulas occurring in the proof are compounded into the 'endsequent', i.e. the formula to be proved.

Those theorems have many interesting applications indicated already by Gentzen (1935), who gave a consistency proof for classical and intuitionistic predicate logic, the solution of the decision problem for intuitionistic propositional logic and a new proof of the nonderivability of the law of the excluded middle in intuitionistic logic. Observe also that the formalism of LK and LJ enables us to characterize very simply classical and intuitionistic logics by restricting the number of 'succedent formulas' in intuitionistic sequents to a single formula.

For classical calculus LK the *Hauptsatz* can be strengthened to the sharpened *Hauptsatz* (Gentzen called it *verschärfter Hauptsatz*; it is known in English also as the midsequent theorem, the normal form theorem, the strengthened *Hauptsatz* and the extended *Hauptsatz*). It says that a proof in LK can be broken up into two parts: one part belonging exclusively to propositional logic, and the other essentially consisting only of application instances of the rules of quantification. This version of the *Hauptsatz* enabled Gentzen to give a new proof of consistency of arithmetic without complete induction (Gentzen 1935, section IV, §3). This is a rather weak form of arithmetic, "of little practical significance" – as Gentzen (1935) wrote. But the problem of consistency of arithmetic with the full induction is much more difficult. Gentzen solved it in two later papers – from 1936 and 1938 – using just the new formalisms introduced in the paper from 1935. His first consistency proof for elementary number theory was formulated in terms of the calculus NK. Gentzen proved the consistency of arithmetic using the transfinite induction up to the ordinal ε_0. The usage of the formalism of natural deduction NK caused the fact that the presentation was neither simple nor elegant. Hence in the later paper from 1938 he used the formalism LK in order to simplify the consistency proof (but this time he lost some of the naturalness in procedure). In a following paper from 1943 Gentzen showed that the transfinite induction up to ε_0 is really necessary in his consistency proof – it cannot be replaced by the induction up to a smaller ordinal $\alpha < \varepsilon_0$.

6.8.5. Those examples indicate already the distinguishing positive marks of the formalisms introduced by Gentzen in comparison with the axiomatic approach. First of all they simulate the natural reasonings, moreover they are an interesting tool which enables us to obtain deep metamathematical results (let us mention here at least the consistency proofs). The formalisms of Gentzen proved to be very flexible – his method is applicable to many non-classical systems, e.g. to the minimal logic (cf. Ketonen 1944), to modal logics (cf. Curry 1952) or even to systems based on certain infinitistic rules of proof (cf. Schütte 1951 where a Gentzen style formalization of an arithmetic based on the ω-rule is given). Gentzen's analysis of the predicate logic in terms of 'natural deduction' influenced also the development of semantic tableaux of Beth (cf. the next section) which are the method for the systematic investigation of the notion of "logical consequence" and the dialogue logic (cf. Lorenzen 1960 and Lorenz 1961).

6.8.6. It is worth noting here some connections between the work of Herbrand and that of Gentzen. Gentzen claimed in (1935) that Herbrand fundamental theorem (cf. previous section) was a special case of his own *verschärfter Hauptsatz* because it applied only to formulas in prenex form. In fact, Herbrand's theorem is more general than Gentzen's and what more the study of Herbrand supplies more information on the midsequent than Gentzen's sharpened *Hauptsatz* is able to provide. Gentzen's *Hauptsatz* on the eliminability of the cut-rule is the analogue of Herbrand's result stating that modus ponens is eliminable in quantification theory. On the other hand, Gentzen's *Hauptsatz* (but not his sharpened *Hauptsatz*) can be extend to various non-classical logics (intuitionistic, modal) while there is no similar extension of Herbrand's result.

Summing up the discussion of Gentzen's work one must admit that Gentzen has in fact presented logic in a fashion more natural than that of Frege, Russell and Hilbert. The formalisms proposed by him have many important features which we tried to indicate above. Admittedly the number of rules in his systems is greater than the number of rules and axioms in, for example, *Principia Mathematica*, but here each sign is introduced separately. The *Hauptsatz* shows an important property of Gentzen's formalisms which has many interesting applications and which leads to many deep mathematical results.

6.9. Semantic and analytic tableaux

6.9.1. The analysis of predicate calculus in terms of natural deduction due to Gentzen has had an influence on the development of logic in another direction. Evert W. Beth (1908–1964) (cf. Beth 1955 and 1959) and independently Jaakko Hintikka (1955) and Kurt Schütte (1956) devised a method dual to the method of Gentzen (1935) called today the tableaux method. As we have shown in the previous section natural deduction was essentially a systematic search for proofs in tree form. The tableaux method is a systematic search for refutations in upside-down tree form. It is based on the construction of a counterexample in cases in which a given formula is not a logical consequence of a given list of formulas. On the other hand it enables us to obtain a cut-free proof of a given true sequent by carring out a thorough, systematic attempt to build a counterexample for it and finding the attempt blocked at some stage from being taken any further.

6.9.2. To describe the tableaux method more exactly we shall present semantic tableaux of Beth (1959). It is based on the following intuition. We want to know if a given formula B is a logical consequence of formulas $A_1, \ldots, A_n$. We test it by constructing a suitable counterexample which should show that it is not the case. If such a counterexample will be found, then we will have a negative answer to our question. And if it will turn out that no suitable counterexample can be found, then we will have an affirmative answer. A systematic method for constructing such counterexamples is provided by drawing up a semantic tableau. We adopt the following rule of building such tableaux:

(A) the tableau consists of two columns – the right and the left (labelled as: valid and invalid, resp.); in the left and in the right column we may insert arbitrary initial formulas,

(B) if the same formula occur in both columns of the same (sub)-tableau, then that (sub)-tableau is closed; if the two subtableaux of a (sub)-tableau are closed, then that (sub)-tableau itself is also closed,

(C) if a formula $\neg A$ occurs in a left column, then A is inserted in the conjugate right column, i.e. in the right column of the same (sub)-tableau and if $\neg A$ occurs in a right column then A is inserted in the conjugate left column,

(D) if a formula $A \& B$ occur in a left column, then both A and B are inserted in the same column; if $A \& B$ appears in a right column, then the (sub)-tableau splits into two sub-tableaux, in the right columns of which we insert, resp., A and B (a sub-tableau is said to be subordinate to those (sub)-tableaux from whose splittings it has arisen; the formulas in both columns of a (sub)-tableau are taken to appear in the corresponding columns of every sub-tableau which is sub-ordinate to it),

(E) if $A \vee B$ occurs in a left column, then the (sub)-tableau splits up into two sub-tableaux, in the left columns of which we insert, resp., A and B; if $A \vee B$ appears in a right column then both A and B are inserted in the same column,

(F) if a formula $A \rightarrow B$ appears in a left column, then the (sub)-tableau splits up: in the right column of one sub-tableau we insert A and in the left column of the other B; if $A \rightarrow B$ occurs in a right column, then B is inserted in the same and B in the conjugate left column,

(G) if a formula $(x)A(x)$ appears in a left column, then we introduce a new same column $A(a)$ for each parameter a which has been or will be introduced; if $(x)A(x)$ appears in a right column, then we introduce a new parameter a and insert $A(a)$ in the same column,

(H) if a formula $(Ex)A(x)$ occurs in a left column, then we introduce a new parameter a and insert $A(a)$ in the same column; if $(Ex)A(x)$ appears in a right column, then we insert in the same column $A(a)$ for each parameter a which has been or will be introduced.

To illustrate those rules we give two examples taken from Beth (1959).

(1) Ask if the formula $(Ez)[P(z)\&\neg S(z)]$ is a logical consequence of formulas: $(Ex)[P(x)\&\neg M(x)]$ and $(Ey)[M(y)\&\neg S(y)]$. According to the above rules we construct the following semantic tableau in which we put the formulas $(Ex)[P(x)\&\neg M(x)]$ and $(Ey)[M(y)\&\neg S(y)]$ in the left column and the formula $(Ez)[P(z)\& \neg S(z)]$ in the right column – this states the conditions to be satisfied by any suitable counterexample.

Valid		Invalid	
(1) $(Ex)[P(x)\&\neg M(x)]$		(3) $(Ez)[P(z)\&\neg S(z)]$	
(2) $(Ey)[M(y)\&\neg S(y)]$		(7) $M(a)$	
(4) $P(a)\&\neg M(a)$		(11) $S(b)$	
(5) $P(a)$		(12) $P(a)\&\neg S(a)$	
(6) $\neg M(a)$		(i)	(ii)
(8) $M(b)\&\neg S(b)$		(13) $P(a)$	(14) $\neg S(a)$
(9) $M(b)$			
(10) $\neg S(b)$		(16) $P(b) \wedge \neg S(b)$	
(i)	(ii)	(iii)	(iv)
	(15) $S(a)$	(17) $P(b)$	(18) $\neg S(b)$
(iii)	(iv)		
	(19) $S(b)$		

Observe that the sub-tableaux (i), (ii) and (iv) are closed and the sub-tableau (iii) is not closed. Hence we obtain a counterexample which can be described as follows: the universe consists of two individuals a and b, the property P belongs to a but not to b, and so does the property S, the property M belongs to b but not to a. In this way we get a negative answer to our initial question.

(2) We ask if the formula $(Ez)[S(z)\&\neg P(z)]$ is a logical consequence of the formulas: $(x)[P(x) \to \neg M(x)]$ and $(Ey)[S(y)\&M(y)]$. We proceed as follows:

Valid		Invalid	
(1) $(x)[P(x) \rightarrow \neg M(x)]$		(3) $(Ez)[S(z)\&\neg P(z)]$	
(2) $(Ey)[S(y)\&M(y)]$		(7) $S(a)\&\neg P(a)$	
(4) $S(a)\&M(a)$			
(5) $S(a)$		(i)	(ii)
(6) $M(a)$		(8) $S(a)$	(9) $\neg P(a)$
(i)	(ii)	(iii)	(iv)
		(12) $P(a)$	(14) $M(a)$
	(10) $P(a)$		
(11) $P(a) \rightarrow \neg M(a)$			
(iii)	(iv)		
	(13) $\neg M(a)$		

Observe that in this tableau all its sub-tableaux are closed, hence the whole tableau is closed. Consequently every model for the formulas (1) and (2) must be a model for the formula (3) and (3) is a logical consequence of the formulas (1) and (2).

6.9.3. The method of semantic tableaux turns out to be equivalent to the natural deduction. In fact, if a semantic tableau does close, i.e. if the construction of a counterexample fails, then one can rearrange it so as to obtain a straightforward derivation, and, conversely, every derivation in a suitable system of natural deduction provides us with a tableau showing that no suitable counterexample can be constructed. Beth proved that his system of tableaux is complete and that it satisfied Gentzen's subformula principle. Consequently, as Beth (1959) writes, "such celebrated and profound results as the theorem of Löwenheim-Skolem-Gödel, the theorem of

Herbrand, or Gentzen's subformula theorem, are from our present standpoint within (relatively) easy reach". And he adds that his approach "realizes to a considerable extend the ideal of a purely analytic method, which has played such an important rôle in philosophy". The method of tableaux has also another philosophical, more exactly epistemological significance, namely it shows in some sense, simply the record of "an unsuccessful attempt to describe counterexample" (Hintikka 1969).

Observe that it cannot be estimated in advance how many steps in the construction of a counterexample by the method of tableaux will be needed, hence the system of the predicate logic is not decidable.

The method of semantic tableaux of Beth was developed by various authors. In particular Kleene (1967) extended it to a Gentzen-type L-system and Kripke (1959, 1959a, 1963) used it in semantics for modal logics. It was applied also in the dialogue logic of Lorenzen (cf. Lorenzen 1960, Lorenz 1961). Smullyan (1968) using the method of Beth and some ideas of Hintikka (1955) introduced the method of analytic tableaux. In some aspects it is similar to the method of semantic tableaux – both try to construct a counterexample. The main difference between them consists in the fact that rules of the analytic tableaux are only rules of elimination of logical constants and that we do not have here two columns of formulas (valid, invalid) but only one in which formulas which in Beth's method would be written in the right column, here stand with negation.

6.9.4. The construction of an analytic tableau can be characterize in the following way. We want to know if the formula B is a consequence of the formulas $A_1, \ldots, A_n$. So write simply formulas $A_1, \ldots, A_n$ and the negation of B, i.e. $\neg B$. Then, applying certain rules, given below, construct a tree which consists of subformulas of $A_1, \ldots, A_n, B$. If every branch of this tree is closed, i.e. if it contains a pair of contradictory formulas, then we conclude that B is a consequence of $A_1, \ldots, A_n$. Otherwise the answer is negative. In each step of the construction of the tree we use that rule which can be applied to the main functor of the formula. The rules are the following:

$(\neg\neg)\quad \dfrac{\neg\neg A}{A}$

$(\&)\quad \dfrac{A\&B}{\begin{array}{c}A\\B\end{array}} \qquad (\neg\&)\quad \dfrac{\neg(A\&B)}{\neg A|\neg B}$

$(\vee)\quad \dfrac{A\vee B}{A|B} \qquad (\neg\vee)\quad \dfrac{\neg(A\vee B)}{\begin{array}{c}\neg A\\\neg B\end{array}}$

$(\rightarrow)\quad \dfrac{A\rightarrow B}{\neg A|B} \qquad (\neg\rightarrow)\quad \dfrac{\neg(A\rightarrow B)}{\begin{array}{c}A\\\neg B\end{array}}$

$(E)\quad \dfrac{(Ex)A(x)}{A(a)} \qquad (\neg E)\quad \dfrac{\neg(Ex)A(x)}{\neg A(a)}$

$(U)\quad \dfrac{(x)A(x)}{A(a)} \qquad (\neg U)\quad \dfrac{\neg(x)A(x)}{\neg A(a)}$

In the case of (E) and $(\neg U)$ we have the following restrictions: the parameter a must not occur earlier in formulas other than A. To give an example ask if the formula $(x)[A(x)\rightarrow B(x)]\rightarrow[(x)A(x)\rightarrow(x)B(x)]$ is a tautology, i.e. a consequence of an empty set of formulas. We construct the following tree:

$$\neg\{(x)[A(x)\rightarrow B(x)]\rightarrow[(x)A(x)\rightarrow(x)B(x)]\}$$
$$(x)[A(x)\rightarrow B(x)]$$
$$\neg[(x)A(x)\rightarrow(x)B(x)]$$
$$(x)A(x)$$
$$\neg(x)B(x)$$
$$A(a)$$
$$\neg B(a)$$
$$A(a)\rightarrow B(a)$$
$$\neg A(a)|B(a)$$

Since both branches of our tree are closed, hence we conclude that our formula is a tautology.

Ask now if a formula $(Ez)[P(z)\&\neg S(z)]$ is a consequence of $(Ex)[P(x)\&\neg M(x)]$ and $(Ey)[M(y)\&\neg S(y)]$. We have the following analytic tableau:

$$
\begin{array}{l}
(Ex)[P(z)\&\neg M(x)] \\
(Ey)[M(y)\&\neg S(y)] \\
\neg(Ez)[P(z)\&\neg S(z)] \\
P(a)\&\neg M(a) \\
P(a) \\
\neg M(a) \\
M(b)\&\neg S(b) \\
M(b) \\
\neg S(b) \\
\neg[P(a)\&\neg S(a)] \\
\quad \begin{array}{c|l}
\neg P(a) & \neg\neg S(a) \\
 & S(a)
\end{array}
\end{array}
$$

We see that the right branch is not closed, hence the answer to our question is negative.

The above examples show that the method of analytic tableaux, which is in a sense of a syntactic character, provides us with a sort of an algorithm – of course it is not an algorithm because the system is undecidable, but in practice it has some features of an algorithm. This method as well as the method of Beth, both based on the idea of decomposition of formulas into simpler components may be treated as realizations of Leibniz's idea of *calculus ratiocinator*.

6.10. Conclusions

We have come to the end of our story of the 20th century way to formalization and mechanization. Its beginnings may be seen in the work of G. Peano who developed the idea of axiomatization of mathematics. The next step formed papers of G. Frege, by whom we find the idea of formalization and in particular the first comprehensive system of logic in which derivations are carried out exclusively according to the form of the expressions. The projects and ideas of Peano and Frege found their fulfilment in the work of B. Russell who realized the idea of formalizing mathematics in his monumental *Principia Mathematica* written together with A. N. Whitehead. This trend was further developed by D. Hilbert and his school. They applied the idea of a formal system to the metamathematical studies of mathematical theories. In this way "the work of Leibniz and Frege has produced results of importance to the progress of mathematics" (Kneale 1962).

One should also mention in this context works of Polish logicians from the so called Lvov-Warsaw logical school, especially J. Łukasiewicz, S. Leśniewski, A. Tarski, B. Sobociński, M. Wajsberg, A. Lindenbaum, S. Jaśkowski and others. They developed first of all formal systems of the propositional calculus and their metatheory paying attention to such properties of systems as independence of axioms or their completeness. They reached really a great formal perfection (for a comprehensive presentation of achievements of Polish logicians of the Lvov-Warsaw school cf. Woleński 1985). And they were conscious of the role they were playing in the process of formalization and mechanization of reasonings. J. Łukasiewicz wrote:

> In Poland the cultivation of mathematical logic has produced more plentiful and fruitful results than in many other countries. We have constructed logical systems which greatly surpass not only traditional logic but also the systems of mathematical logic formulated until now. We have understood, perhaps better than others, what a deductive system is and how such systems should be built. We have been the first to grasp the connexion of mathematical logic with the ancient systems of formal logic. Above all, we have achieved standards of scientific precision that are much superior to the requirements accepted so far.
>
> Compared with these new standards of precision, the exactness of mathematics, previously regarded as an unequalled model, has not held its own. The degree of precision sufficient for the mathematician does not satisfy us any longer. We require that every branch of mathematics should be a correctly constructed deductive system. We want to know the

> axioms on which each system it based, and the rules of inference of which it makes use. We demand that proofs should be carried out in accordance with these rules of inference, that they should be complete and capable of being mechanically checked. We are no longer satisfied with ordinary mathematical deductions, which usually start somewhere 'in the middle', reveal frequent gaps, and constantly appeal to intuition (cf. Łukasiewicz 1961, see also Łukasiewicz 1970, p.111).

New direction in researches of formalization and mechanization of reasonings was suggested by L. Löwenheim and T. Skolem who attempt to analyse quantified statements in terms of truth-functional logic. It was further developed by J. Herbrand. Another new approach was introduced by G. Gentzen who broke away from the traditional formulations of predicate logic as they were developed by Frege, Russell and Hilbert and presented the systems of natural deduction which are closer to the real reasonings and which enable us to prove various interesting metamathematical results.

The last step on the 20th century way to formalization and mechanization we indicated was the method of tableaux of Beth and Smullyan which is in fact a method of search for refutations and in this way provides us with a sort of an algorithm of proving formulas.

One should add to all that still one minor remark concerning the mechanization of the process of verification if a given formula of the propositional calculus is a tautology or not. We can find already in Boole's work the notion that a necessarily true proposition should reveal itself as such in the process of development. About the year 1920 it occurred to several logicians independently that when truth-function had been defined by means of truth tables a rule of thumb calculation would suffice to determine whether a complex proposition expressed by the use of truth-functional signs was necessarily true (i.e. a tautology), necessarily false (i.e. a contratautology), or contingent. The best known account of such a decision procedure is that given by Wittgenstein (1922) (a more straightforward discussion of the technique is in Post 1921; note that the method was also known to Łukasiewicz at this time).

All those achievements made it possible to develop the mechanization of reasonings. The artificial formal languages and the idea of formalization afforded possibilities for representing appropriately thoughts and for grasping and expressing properly the fundamental relations in reasonings. The methods developed by Skolem, Herbrand, Gentzen and Beth provided procedures which enabled the simplification of reasonings and further their mechanizations, for a reasoning is, after all, a kind of computation. Those methods made it possible to construct concrete computer programs which could prove some theorems of, say, predicate calculus. For example, P. C. Gilmore using Beth's semantic tableaux technique constructed the first working mechanized proof procedure for the predicate calculus and did prove some theorems of modest difficulty (cf. Gilmore 1959 and 1960). Hao Wang developed a program which proved 220 theorems of *Principia Mathematica* (in 37 minutes) – the program was organized around Gentzen-Herbrand methods (cf. Wang 1960, 1960a, 1961). These were the beginnings of the study of the automated theorem proving which is developed nowadays with great intensity. We discuss those and other related topics in the next chapter.

CHAPTER SEVEN

MECHANIZED DEDUCTION SYSTEMS

7.1. Introduction

7.1.1. In 1936 Alan Turing and Alonzo Church proved two theorems which seemed to have destroyed all hopes of establishing a method of mechanizing reasonings. Turing in (1936–37) reduced the decidability problem for theories to the halting problem for abstract machines modelling the computability processes (and named after him) and proved that the latter is undecidable. Church (1936) solving Hilbert's original problem proved the undecidability of the full predicate logic and of various subclasses of it.

On the other hand results of Skolem and Herbrand (cf. the previous chapter) showed that if a theorem is true then this fact can be proved in a finite number of steps – but this is not the case if the theorem is not true (in this situation either one can prove in some cases the falsity of the given statement or the verification procedure does not halt). This semidecidability of the predicate logic was the source of hope and the basis of further searches for the mechanized deduction systems. Those studies were heavily stimulated by the appearence of computers in early fifties. There appeared the idea of applying them to the automatization of logic by using the mechanization procedures developed earlier. The appearance of computers stimulated also the search for new, more effecitive procedures.

7.1.2. In the sequel we shall describe the history of those researches. In Section 7.2 the early attempts of applying computers to prove theorems will be presented, in particular we shall tell about results of Davis, Newell-Shaw-Simon, Gilmore, Gelernter *et al.*, Hao Wang and Davis-Putnam. Section 7.3 will be devoted to resolution and unification algorithms of Prawitz and Robinson and to their modifications. They turned out to be crucial for the further development of the researches towards mechanization and automatization of reasonings. We will sketch them in Section 7.4.

7.1.3. All those studies led to the idea of automated theorem proving by which one means the use of a computer to prove non-numerical results, i.e. determine their truth (validity). One can demand here either a simple statement "proved" (what is the case in decision procedures) or human readable proofs. We can distinguish also two modes of operation: 1° fully automated proof searches and 2° man-machine interaction proof searches.

7.1.4. Note that the studies of mechanization of reasonings and automated theorem proving were motivated by two different philosophies. The first one – which we shall call the logic approach – can be characterized "by the presence of a dominant logical system that is carefully delineated and essentially static over the development stage of the theorem proving system. Also, search control is clearly separable from the logic and can be regarded as sitting 'over' the logic. Control 'Heuristics' exist but are syntax-directed primarily" (cf. Loveland 1984, p.3). The second philosophical viewpoint is called the human simulation approach. It is generally the antithesis of the first one. It can be characterized shortly by saying that "the thrust is obviously simulation of human problem solving techniques" (cf. Loveland 1984, p.3). Of course the logic and human simulation approaches are not always clearly delineated. Nevertheless this distinction will help us to order the results and to consider the influence of each of the approaches on various particular systems.

7.2. First mechanized deduction systems

7.2.1. The different philosophical backgrounds mentioned above can be spotted already in the first studies towards mechanization of reasonings and automated theorem proving. In 1954 Martin Davis wrote a program to prove theorems in additive arithmetic (this program was never published in a paper). It was developed on the computer "Johniac" in the Institute for Advance Studies and was a straight implementation of the classical Presburger's decision procedure for additive number theory (i.e. for the theory of non-negative integers in the language with zero, successor and addition only) (cf. Presburger 1929). The complexity of this decision procedure is very high and therefore the program proved only very simple facts (e.g. that the sum of two even numbers is also an even number – this was the first mathematical theorem in the history proved by a computer!).

7.2.2. The Presburger prover of M. Davis was an example of the logic approach. The second achievement in the field of automated theorem proving called the logic theorist should be included among examples of the human simulation approach. We mean here the program of A. Newell, J. S. Shaw and H. A. Simon presented in 1956 at the Dartmouth conference (cf. Newell *et al.* 1956). This program could prove some theorems in the propositional calculus of *Principia Mathematica* of A. N. Whitehead and B. Russell. Its goal was to mechanically simulate the deduction processes of humans as they prove theorems of the sentential calculus. Two methods were used: 1° substitution into established formulas to directly obtain a desired result and, failing that, 2° to find a subproblem whose proof represents progress towards proving the goal problem. This program was able to prove 38 theorems of *Principia.*

7.2.3. The Geometry Theorem-proving Machine of Gelernter and others from 1959 (cf. Gelernter *et al.* 1959, 1960) is also an example of the second approach. It applied the idea of M. Minsky that the diagram that traditionally accompanies plane geometry problems is a simple model for the theorem that could greatly prune the proof search. The program worked backwards, i.e. from the conclusion (goal) towards the premises creating new subgoals. The geometry model was used just to say which subgoals were true and enabled to drop the false ones. It should be noticed that this program was able to prove most high school exam problems within its domain and the running time was often comparable to high school student time.

7.2.4. Simultaneously with the Geometry Theorem-proving Machine two new efforts in the logic framework occurred. We mean here works of Gilmore and Hao Wang. They used methods derived from classical logic proof procedures and in this way rejected opinions that logical methods cannot provide a useful basis for automated theorem proving. Such opinions were rather popular at that time. They were founded on the fact that logic oriented methods were inefficient and on the fact that the methods of Newell, Shaw, Simon and Gelernter proved to be successful. The method of P. C. Gilmore was based on Beth's semantic tableau technique (cf. the previous chapter). It was probably the first working mechanized proof procedure for the predicate logic – it proved some theorems of modest difficulty (cf. Gilmore 1959, 1960).

7.2.5. In the summer 1958 Hao Wang developed the first logic oriented program of automated theorem proving of IBM and continued this work at Bell Labs in 1959–63 (cf. Wang 1960, 1960a, 1961). Three programs were developed: 1° for propositional calculus, 2° for a decidable part of the predicate calculus and 3° for all of predicate calculus. Those programs were based on Gentzen-Herbrand methods (cf. the previous chapter), the last one proved about 350 theorems of *Principia Mathematica* (they were rather simple theorems of pure predicate calculus with identity).

7.2.6. During the Summer Institute for Symbolic Logic held at Cornell University, USA, in 1954 Abraham Robinson put forward, in his short lecture, the idea of considering the additional points, lines and circles – which must be used in a search for a solution of a geometrical problem – simply as elements of the so called Herbrand universe. This should enable us to abandon the geometrical constructs and to use directly Herbrand's methods.

7.2.7. This idea turned out to be very influential and significant. One of the first programs that realized it was implemented in 1960–62 by M. Davis and H. Putnam (cf. Davis–Putnam 1960). By Herbrand's theorem – as was shown in the previous chapter – the question of validity of a predicate calculus formula Z can be reduced to a series of validity questions about ever-expanding propositional formulas. More exactly one should consider so called Herbrand disjunctions $A_1 \vee \cdots \vee A_n$ (which can be effectively obtained from Z). It holds that Z is valid if and only if there exists an n such that the disjunction $A_1 \vee \cdots \vee A_n$ is valid. The formulas A_i are essentially substitition instances over an expanded term alphabet of Z with quantifiers removed. So one can test now $A_1 \vee \cdots \vee A_n$ for ever-increasing n for example by truth table and conclude that Z is valid if among formulas $A_1 \vee \cdots \vee A_n$ a tautology was found. But truth tables happen to be quite inefficient. The procedure of Davis and Putnam tried to overcome this difficulty. In fact they were considering unsatisfiability (instead of validity) of formulas and worked with conjunctive normal forms. Such a form is a conjunction of clauses, each clause being a disjunction of literals, i.e. atomic formulas (atoms) or their negations. The Davis–Putnam procedure can be described now as follows: "[it] made optimal use of simplification by cancellation due to one-literal clauses or because some literals might not have their complement (the same atomic formula but only one with a negation sign) in the formula. A simplified formula was split into two formulas so further simplification could recur anew" (cf. Loveland 1984).

7.3. Unification and resolution

7.3.1. The procedure of Davis and Putnam described in the last section had some defects. The main one was the enumeration substitutions – prior to this point substitutions were determined by some enumeration scheme that covered every possibility. Prawitz (1960) realized that the only important substitutions create complementary literals. He found substitutions by deriving a set of identity conditions (equations) that will lead to contradictory propositional formula if the conditions are met. In this way one got a procedure of substituting Herbrand terms. It is called today unification.

7.3.2. M. Davis developed right away the idea and combining it with the procedure of Davis–Putnam implemented it in a computer program based on a so called Linked-Conjunct method (cf. Davis 1963). This program used conjunctive normal forms of formulas and the unification algorithm developed by D. McIlroy in November 1962 at the Bell-Telephone-Laboratories. It was the first program which overcame the weaknesses of Herbrand procedure and improved it just by using conjunctive normal form and the unification algorithm (cf. Davis 1963 and Chinlund *et al.* 1964).

7.3.3. Simultaneously at Argonne National Lab near Chicago a group of scientists (G. Robinson, D. Carson, J. A. Robinson, L. Wos) was working on computer programs proving theorems. They used methods based on Herbrand theorem recognizing their inefficiency. Studying papers of Davis–Putnam and Prawitz they came to the idea of trying to find a general machine-oriented logical principle which would unify their ideas in a single rule of inference. Such a rule was found in 1963–64 by John Alan Robinson and published in (1965). It is called the resolution principle and is today one of the most fundamental ideas in the field of automated theorem proving. Therefore we shall describe it now more exactly.

7.3.4. The resolution principle is applied to formulas in a special form called conjunctive normal form. Given a formula A of the language of predicate calculus we first transform it into prenex normal form, i.e. to the form $(Q_1x_1)(Q_2x_2)\ldots(Q_nx_n)B$ where Q_i $(i = 1,\ldots,n)$ are universal or existential quantifiers and B is quantifier-free (B is called matrix). One can show that the prenex normal form of a formula A is logically equivalent to the formula A. As an example consider the following formula (stating that the function f is continuous):

$$\forall\varepsilon\{\varepsilon > 0 \longrightarrow \exists\delta[\delta > 0 \wedge \forall x\forall y(|x-y| < \delta \longrightarrow |f(x)-f(y)| < \varepsilon)]\}.$$

Its prenex normal form is:

$$\forall\varepsilon\exists\delta\forall x\forall y[\varepsilon > 0 \longrightarrow \delta > 0 \wedge (|x-y| < \delta \longrightarrow |f(x)-f(y)| < \varepsilon)].$$

Observe that prenex normal form of a formula is not uniquely determined but on the other hand all prenex normal forms of a given formula are logically equivalent.

7.3.5. The next step of the transformation of formulas we need is skolemization. We described it already in Section 6.5 if the previous chapter. Recall here only that it consists of dropping of all the quantifiers and of replacing every occurrence of each variable bound by an existential quantifier by a functional term containing the variables that in this formula are bound by universal quantifiers preceding the considered existential quantifier. A formula obtained in this way is said to be in a skolemized prenex normal form or in its functional form for satisfiability. The skolemized prenex normal form of the formula from our above example is the following:

$$\varepsilon > 0 \longrightarrow g(\varepsilon) > 0 \wedge (|x - y| < g(\varepsilon) \longrightarrow |f(x) - f(y)| < \varepsilon).$$

Note that the skolemized normal form of a formula is not equivalent to the given formula but it is satisfiable (inconsistent) if and only if the given formula is satisfiable (inconsistent).

7.3.6. The next step of getting the conjunctive normal form consists of the elimination of connectives $\longleftrightarrow$ and $\longrightarrow$ and moving all negation signs $\neg$ to the atoms. We proceed according to the following rules:

$$\begin{aligned}
A \longleftrightarrow B &\text{ is logically equivalent to } (A \longrightarrow B) \wedge (B \longrightarrow A)\\
A \longrightarrow B &\text{ is logically equivalent to } (\neg A \vee B)\\
\neg(A \wedge B) &\text{ is logically equivalent to } (\neg A \vee \neg B)\\
\neg(A \vee B) &\text{ is logically equivalent to } (\neg A \wedge \neg B)\\
\neg\neg A &\text{ is logically equivalent to } A.
\end{aligned}$$

A formula obtained in such a way is said to be in the negation normal form. For example the negation normal form of the formula considered above is:

$$\neg \varepsilon > 0 \vee [g(\varepsilon) > 0 \wedge (\neg |x - y| < g(\varepsilon) \vee |f(x) - f(y)| < \varepsilon)].$$

Note that atoms and negated atoms are usually called literals.

The last step of the considered transformation consists of the application of the following distributivity laws:

$$\begin{aligned}
A \vee (B \wedge C) &\longleftrightarrow (A \vee B) \wedge (A \vee C),\\
(A \wedge B) \vee C &\longleftrightarrow (A \vee C) \wedge (B \vee C).
\end{aligned}$$

In this way the obtained formula is of the form of conjunctions of disjunctions of literals. Such a form is called the conjunctive normal form and the disjunctions of literals are called clauses. Clauses consisting of a single literal are called unit clauses. Clauses with only one positive literal are called Horn clauses. They correspond to formulas of the form $A_1 \wedge \ldots \wedge A_n \longrightarrow B$. To illustrate the last step note that the conjunctive normal form of our formula stating that a function f is continuous is the following:

$$(\neg\varepsilon > 0)\vee[g(\varepsilon) > 0\wedge\neg|x-y| < g(\varepsilon)]\vee[g(\varepsilon) > 0\wedge|f(x)-f(y)| < \varepsilon].$$

Note that if $(A_1^1\vee\cdots\vee A_{n_1}^1)\wedge\ldots\wedge(A_1^l\vee\cdots\vee A_{n_l}^l)$ is a conjunctive normal form of a formula A then we write it sometimes also as:
$\{A_1^1\vee\cdots\vee A_{n_1}^1,\ldots,A_1^l\vee\cdots\vee A_{n_l}^l\}$

or as

$\{\{A_1^1,\cdots,A_{n_1}^1\},\ldots,\{A_1^l,\cdots,A_{n_l}^l\}\}$.
This form is called the clause form. In this way we have shown that for any formula A of the language of predicate calculus there exists a formula A' in conjunctive normal form such that the formula A is satisfiable (inconsistent) if and only if the formula A' is satisfiable (inconsistent).

7.3.7. Having described the needed form of formulas we can introduce now the resolution principle. First observe that a formula B is a logical consequence of formulas $A_1,\ldots,A_n$ if and only if the formula $A_1\wedge\ldots\wedge A_n\wedge\neg B$ is inconsistent, i.e. unsatisfiable. Let $\Box$ denote an empty clause (i.e. a contradiction) and let formulas $A_1,\ldots,A_n,\neg B$ be in conjunctive normal form. Hence to show that B follows logically from $A_1,\ldots,A_n$ it suffices to prove that $\Box$ is contained in the set $\mathcal{S}$ of all clauses constituting $A_1,\ldots,A_n,\neg B$ or that $\Box$ can be deduced from this set (the sense of the word 'deduce' will be explained below). This is the main idea of the method of resolution. Hence we can say that this method is a negative test calculus.

The resolution calculus introduced by J. A. Robinson (1965) is a logical calculus in which one works only with formulas in clause form. It has no logical axioms and only one inference rule (the resolution rule). The simplest version of this rule has the following form: if C_1 and C_2 are two clauses such that C_1 contains a literal L_1 and C_2 contains a literal L_2 which is inconsistent with L_1 (i.e. L_1 and L_2 are complementary literals) then one obtains a new clause C consisting of all literals of C_1 except L_1 and all literals of C_2 except L_2. Symbolically it can be written as:

$$\begin{array}{rl}\text{clause } C_1: & K_1\vee\cdots\vee K_n\vee L\\ \text{clause } C_2: & M_1\vee\cdots\vee M_m\vee\neg L\\ \hline & K_1\vee\cdots\vee K_n\vee M_1\vee\cdots\vee M_m\end{array}$$

The clause $K_1 \vee \cdots \vee K_n \vee M_1 \vee \cdots \vee M_m$ is called the resolvent of C_1 and C_2 and clauses C_1 and C_2 are called parent clauses. We say that L_1 and L_2 are the literals resolved upon when the resolvent exists. Observe that the resolvent is a logical consequence of the parent clauses.

Let $\mathcal{S}$ be a set of clauses. A refutation (or a proof of unsatisfiability) of $\mathcal{S}$ is a finite sequence $C_1, \ldots, C_n$ of clauses such that 1° any C_i $(i = 1, \ldots, n)$ either belongs to $\mathcal{S}$ or there exist C_j and C_k, $j, k < i$ such that C_i is a resolvent of C_j and C_k and 2° the last clause C_n is $\Box$. This explains what we meant above by saying that $\Box$ can be "deduced" from the set $\mathcal{S}$ of clauses.

We give now some examples.

A. We show that U is a logical consequence of formulas $P \longrightarrow S$, $S \longrightarrow U$ and P. Indeed it suffices to show that the formula $(P \longrightarrow S) \wedge (S \longrightarrow U) \wedge P \wedge \neg U$ is unsatisfiable (inconsistent). Writing it in conjunctive normal form we get

$$(\neg P \vee S) \wedge (\neg S \vee U) \wedge P \wedge \neg U.$$

Its clause form is $\{\neg P \vee S, \neg S \vee U, P, \neg U\}$. Denote this set of clauses by $\mathcal{S}$. The following sequence of formulas is a proof of unsatisfiability of $\mathcal{S}$:

(1) $\neg P \vee S$
(2) $\neg S \vee U$
(3) P
(4) $\neg U$
(5) $\neg P \vee U$ resolvent of (1) and (2)
(6) U resolvent of (3) and (5)
(7) $\Box$ resolvent of (4) and (6).

Hence U is a logical consequence of $P \longrightarrow S$, $S \longrightarrow U$ and P.

B. We show that the formula $\neg Q$ is a logical consequence of $P \longrightarrow (\neg Q \vee (R \wedge S)), P, \neg S$. Hence consider the formula

$$(P \longrightarrow (\neg Q \vee (R \wedge S))) \wedge P \wedge \neg S \wedge Q.$$

Its clause form is

$$\mathcal{S} = \{\neg P \vee \neg Q \vee R, \neg P \vee \neg Q \vee S, P, \neg S, Q\}.$$

We have the following proof of unsatisfiability of $\mathcal{S}$:

(1) $\neg P \vee \neg Q \vee R$
(2) $\neg P \vee \neg Q \vee S$
(3) P
(4) $\neg S$
(5) Q
(6) $\neg P \vee \neg Q$ resolvent of (2) and (4)
(7) $\neg P$ resolvent of (5) and (6)
(8) $\square$ resolvent of (3) and (7).

7.3.8. So far we considered the simplest form of the resolution principle and its application in the propositional calculus. In the case of formulas containing variables the whole situation is more complicated.

First we describe a substitution device called unification (we shall do it following Chang-Lee 1973). By a substitution we mean a finite set of the form $\{t_1/v_1, \ldots, t_n/v_n\}$ where v_i are variables and t_i are terms different from v_i. An empty substitution will be denoted by ε. If $\theta = \{t_1/v_1, \ldots, t_n/v_n\}$ is a substitution and E is a formula then by $E\theta$ we denote the formula $E(v_1/t_1, \ldots, v_n/t_n)$. Observe that substitutions can be composed, i.e. if $\theta = \{t_1/x_1, \ldots, t_n/x_n\}$ and $\lambda = \{u_1/y_1, \ldots, u_m/y_m\}$ are two substitutions then by $\theta \circ \lambda$ we shall denote a composition of θ and λ and define it as a substitution obtained from the set $\{t_1\lambda/x_1, \ldots, t_n\lambda/x_n, u_1/y_1, \ldots, u_m/y_m\}$ by removing from it all the elements $t_j\lambda/x_j$ such that $t_j\lambda = x_j$ and all the elements u_i/y_i such that $y_i \in \{x_1, \ldots, x_n\}$. Note that the composition $\circ$ is associative and that ε is the left and right unit, i.e. $\varepsilon \circ \theta = \theta \circ \varepsilon = \varepsilon$.

A substitution θ is said to be a unifier of a set of formulas $\{E_1, \ldots, E_k\}$ if and only if $E_1\theta = E_2\theta = \cdots = E_k\theta$. If there exists a unifier of a set $\{E_1, \ldots, E_k\}$ then this set is said to be unifiable. A unifier σ of the set $\{E_1, \ldots, E_k\}$ is called a most general unifier if and only if for any unifier θ of this set there exists a substitution λ such that $\theta = \sigma \circ \lambda$. J. A. Robinson showed that for any set $\mathcal{S}$ of formulas there exists at most one most general unifier.

We shall describe now an algorithm of finding a most general unifier. It is called the unification algorithm.

Let $\mathcal{S}$ be a nonempty set of formulas. A disagreement set of $\mathcal{S}$ is defined as follows: one indicates the first (from the left) position such that there are two formulas from $\mathcal{S}$ which differ on this position. Then for every element E of $\mathcal{S}$ we write its subformula beginning with the symbol being written on this position. The set of these subformulas is just the disagreement set of $\mathcal{S}$. For example the disagreement set of the set $\{P(x, f(y, z)), P(x, a), P(x, g(h(k(x))))\}$ is $\{f(y, z), a, g(h(k(x)))\}$.

The unification algorithm can now be given by the following rules. Let a set $\mathcal{S}$ of formulas be given.

Step 1: $\mathcal{S}_0 = \mathcal{S}$, $\sigma_0 = \varepsilon$.

Step 2: If $\mathcal{S}_k$ is a unit clause then STOP and σ_k is the most general unifier of $\mathcal{S}$. Otherwise let D_k be the disagreement set of $\mathcal{S}_k$.

Step 3: If there are v_k and t_k in D_k such that v_k is a variable not occuring in t_k then move to Step 4. Otherwise STOP: $\mathcal{S}$ is not unifiable.

Step 4: Let $\sigma_{k+1} = \sigma_k \circ \{t_k/v_k\}$ and $\mathcal{S}_{k+1} = \mathcal{S}_k\{t_k/v_k\}$. (Observe that $\mathcal{S}_{k+1} = \mathcal{S}\sigma_{k+1}$.)

Note that the unification algorithm always halts when applied to a finite nonempty set of formulas. J. A. Robinson proved that if $\mathcal{S}$ is a finite nonempty unifiable set of formulas then the unification algorithm halts on Step 2 and the last σ_k is the most general unifier of $\mathcal{S}$.

We give some examples. First find a most general unifier of the set $\mathcal{S} = \{Q(f(a), g(x)), Q(y, y)\}$. We proceed as follows:

1. $\sigma_0 = \varepsilon$, $\mathcal{S}_0 = \mathcal{S}$.

2. Since $\mathcal{S}_0$ is not a unit clause we find a disagreement set D_0 of $\mathcal{S}_0$. We have $D_0 = \{f(a), y\}$. Hence $v_0 = y, t_0 = f(a)$.

3. $\sigma_1 = \sigma_0 \circ \{t_0/v_0\} = \sigma_0 \circ \{f(a)/y\} = \{f(a)/y\}$,
$\mathcal{S}_1 = \mathcal{S}_0\{t_0/v_0\} = \{Q(f(a), g(x)), Q(f(a), f(a))\}$.

4. $\mathcal{S}_1$ is not a unit clause and we have $D_1 = \{g(x), f(a)\}$. By Step 3 we conclude that the set $\mathcal{S}$ is not unifiable.

We give one more example.

Let $\mathcal{S} = \{P(a, x, f(g(x))), P(z, f(z), f(u))\}$. Find a most general unifier of $\mathcal{S}$. We proceed as follows:

1. $\sigma_0 = \varepsilon$, $\mathcal{S}_0 = \mathcal{S}$.

2. $D_0 = \{a, z\}$ and we have $v_0 = z$, $t_0 = a$.

3. $\sigma_1 = \sigma_0 \circ \{t_0/v_0\} = \{a/z\}$,
   $\mathcal{S}_1 = \mathcal{S}_0\{t_0/v_0\} = \{P(a, x, f(g(y))), P(a, f(a), f(u))\}$.

4. $D_1 = \{x, f(a)\}$ and we put $v_1 = x$, $t_1 = f(a)$.

5. $\sigma_2 = \sigma_1 \circ \{t_1/v_1\} = \{a/z, f(a)/x\}$,
   $\mathcal{S}_2 = \mathcal{S}_1 \circ \{t_1/v_1\} = \{P(a, f(a), f(g(y))), P(a, f(a), f(u))\}$.

6. $D_2 = \{g(y), u\}$ and we put $v_2 = u$, $t_2 = g(y)$.

7. $\sigma_3 = \sigma_2 \circ \{t_2/v_2\} = \{a/z, f(a)/x, g(y)/u\}$,
   $\mathcal{S}_3 = \mathcal{S}_2\{t_2/v_2\} = \{P(a, f(a), f(g(y)))\}$.

By Step 2 the set $\mathcal{S}$ is unifiable and σ_3 is the most general unifier of it.

7.3.9. We have to define two more notions to formulate at last the general form of the resolution rule. If two or more literals (with the same sign) of a clause C have a most general unifier σ then the clause $C\sigma$ is called a factor of C. Let C_1 and C_2 be two clauses which have no common variable and let L_1 and L_2 be two literals of C_1 and C_2, resp. If L_1 and $\neg L_2$ have a most general unifier σ then the clause $(C_1\sigma - L_1\sigma) \cup (C_2\sigma - L_2\sigma)$ is called a binary resolvent of C_1 and C_2. The clauses C_1 and C_2 are called parent clauses and we say that L_1 and L_2 are literals resolved upon. A resolvent of two parent clauses C_1 and C_2 is now defined as one of the following resolvents: (1) binary resolvent of C_1 and C_2, (2) binary resolvent of C_1 and a factor of C_2, (3) binary resolvent of C_2 and a factor of C_1, (4) binary resolvent of a factor C_1 and a factor of C_2.

Now we can define a resolution deduction. Let a set $\mathcal{S}$ of clauses and a clause A be given. A resolution deduction of A from $\mathcal{S}$ is a finite sequence $C_1, \ldots, C_n$ of clauses such that: (1) C_n is A, (2) for any $i, 1 \leq i \leq n, C_i$ is either a member of $\mathcal{S}$ or there exist $j, k < i$ such that C_i is a resolvent of C_j and C_k (i.e. C_i is obtained from C_j and C_k by the resolution rule). A resolution deduction of the empty clause $\Box$ from $\mathcal{S}$ is called a refutation (or a proof of unsatisfiability) of $\mathcal{S}$.

We shall give some examples (cf. Chang-Lee 1973).

1. Show that the formula $B =: \exists x(S(x) \wedge R(x))$ is a logical consequence of formulas $A_1 =: \forall x[P(x) \longrightarrow (Q(x) \wedge R(x))]$ and $A_2 =: \exists x(P(x) \wedge S(x))$. It suffices to show that the formula $A_1 \wedge A_2 \wedge \neg B$ is unsatisfiable. Transform the given formulas $A_1, A_2, \neg B$ into conjunctive normal form. We get the following formulas, resp.,

$$(\neg P(x) \vee Q(x)) \wedge (\neg P(x) \vee R(x)),$$
$$P(a) \wedge S(a),$$
$$\neg S(x) \vee \neg R(x).$$

We can construct now the following resolution deduction:

(1)	$\neg P(x) \vee Q(x)$	from A_1
(2)	$\neg P(x) \vee R(x)$	
(3)	$P(a)$	from A_2
(4)	$S(a)$	
(5)	$\neg S(x) \vee \neg R(x)$	from $\neg B$
(6)	$R(a)$	resolvent of (2) and (3)
(7)	$\neg R(a)$	resolvent of (4) and (5)
(8)	$\square$	resolvent of (6) and (7).

Hence we have shown that B is a logical consequence of A_1 and A_2.

2. We show that the formula $B =: \exists x(P(x) \wedge R(x))$ is a logical consequence of the following formulas:

$$A_1 =: \forall x[Q(x) \wedge \neg T(x) \longrightarrow \exists y(S(x,y) \wedge R(y))],$$
$$A_2 =: \exists x[P(x) \wedge Q(x) \wedge \forall y(S(x,y) \longrightarrow P(y))],$$
$$A_3 =: \forall x(P(x) \longrightarrow \neg T(x)).$$

Transforming the formulas A_1, A_2, A_3 and $\neg B$ into conjunctive normal form we get the following clauses:

(1)	$\neg Q(x) \vee T(x) \vee S(x, f(x))$	from A_1
(2)	$\neg Q(x) \vee T(x) \vee R(f(x))$	
(3)	$P(a)$	from A_2
(4)	$Q(a)$	
(5)	$\neg S(a,y) \vee P(y)$	
(6)	$\neg P(x) \vee \neg T(x)$	from A_3
(7)	$\neg P(x) \vee \neg R(x)$	from $\neg B$

The needed deduction of □ can be the following:

(8)	$\neg T(a)$	resolvent of (3) and (6)
(9)	$T(a) \vee R(f(a))$	resolvent of (2) and (4)
(10)	$R(f(a))$	resolvent of (8) and (9)
(11)	$T(a) \vee S(a, f(a))$	resolvent of (1) and (4)
(12)	$S(a, f(a))$	resolvent of (8) and (11)
(13)	$P(f(a))$	resolvent of (5) and (12)
(14)	$\neg R(f(a))$	resolvent of (7) and (13)
(15)	□	resolvent of (10) and (14).

7.3.10. The method of resolution has an important property called the refutation completeness. One can prove that a set $\mathcal{S}$ of clauses is unsatisfiable if and only if there exists a resolution deduction of the empty clause □ from $\mathcal{S}$. The proof of this fact uses semantical trees and Herbrand's theorem. Hence the method gives a semidecidability of the predicate logic: if a given formula is a logical consequence of a given set $\mathcal{S}$ of formulas then by a systematic application of the resolution rule we get in a finite number of steps the empty clause □, but if it is not a logical consequence then sometimes one can decide it after a finite number of steps but in general the procedure does not halt. On the other hand the considered method does not give a procedure of finding a formal proof of a formula B on the basis of formulas $A_1, \ldots, A_n$ in the case when B is a logical consequence of $A_1, \ldots, A_n$.

7.3.11. Recall that we used in the resolution most general unifiers, i.e. most general substitutions that allow the equality of literals. This guarantees the elimination of branching of search due to different possible substitutions that equate those atoms but lead to different clauses. Therefore the method of resolution is simple, elegant and powerful. But the world is not so perfectly beautiful – this method has also some defects. If one generates from a given set of clauses new clauses by the method of resolution then they accumulate at a rapid rate. Indeed given a set $\mathcal{S}$ of clauses one can obtain new clauses systematically using the level-saturation method which is described by the following equations: $\mathcal{S}^0 = \mathcal{S}, \mathcal{S}^{n+1} = \{$resolvents of C_1 and C_2: $C_1 \in \mathcal{S}^0 \cup \ldots \cup \mathcal{S}^n, C_2 \in \mathcal{S}^n\}\}, n = 1, 2, \cdots$ In this way we get all the resolvents of all pairs of elements of $\mathcal{S}$, add them to $\mathcal{S}$, further we calculate all the resolvents of elements of this new set etc. till we come to the empty clause □. Among clauses generated in this procedure there are many irrelevant ones and the total number of clauses grows rapidly. Hence the idea of improving the method of resolution by finding restrictions and strategies to control the growth of the number of clauses. We shall tell here only about some of the proposed improvements (Loveland 1978 summarizes in Appendix twenty five such improvements but more exist). By a restriction of resolution we mean a variant for which some clauses generated by the basic resolution procedure are not generated. A strategy of resolution only rearranges the order of generation to get likely useful clauses earlier.

7.3.11.1. One of the earliest refinements of resolution is unit-preference introduced in Wos-Carson-Robinson (1964). This strategy guarantees that the resolvent is shorter than the longer parent clauses. In the same paper L. Wos, D. F. Robinson and G. A. Carson introduced the set-of-support restriction. It can be described as follows: one chooses a subset $\mathcal{T}$ (called a support) of a given set $\mathcal{S}$ of clauses and then two clauses from $\mathcal{S} - \mathcal{T}$ are never resolved together. This means that every resolvent has in its deduction history some clause of $\mathcal{T}$. In practice $\mathcal{T}$ is usually chosen to be a (sub)set of the clauses special to the considered problem.

7.3.11.2. J. A. Robinson in (1965a) introduced a restriction called hyperresolution. It restricts resolutions to where one parent clause contains only positive literals and any resolvent containing a negative literal is immediately used in all permitted resolutions and then discarded.

7.3.11.3. D. W. Loveland (1970) and D. Luckham (1970) proposed another restriction called linear resolution (Luckham used the name "ancestry-filter form"). It constrains the deduction so that a new clause is always derived from the preceding clause of the deduction by resolving against an earlier clause of the deduction. In this way one is always seeking to transform the last clause obtained into a clause closer to the goal clause. This method was further developed by R. Anderson and W. W. Bledsoe (1970). R. Yates, B. Raphael and T. Hart (1970), R. Reiter (1971), D. W. Loveland (1972) and R. Kowalski and D. Kuehner (1971).

7.3.11.4. D. W. Loveland (1968) and (1969) introduced a procedure which is not a resolution procedure but is close to a very restricted form of linear resolution – it is called model elimination. R. Kowalski and D. Kuehner (1971) provided the translation of it into a very restricted linear resolution format called SL-resolution. It was used by A. Colmerauer and P. Roussel to an early version of a programming language Prolog.

7.3.11.5. There appeared a number of resolution refinements which reduce multiple derivations of the same clause by ordering literals in clauses. An example of this type of procedures is the method called locking or lock-resolution introduced by R. S. Boyer (1971). Its main idea is to use indices to order literals in clauses from a given set of clauses. The occurrences of literals are indexed by integers. Then resolution need be done using only the lowest indexed integer of each clause.

7.3.11.6. We should mention here also the semantic resolution of J. R. Slagle (1967) which generalizes hyperresolution of J. A. Robinson (1965a), resolution with renaming of B. Meltzer (1966) and the set-of-support restriction of L. Wos, G. A. Robinson and D. F. Carson (1965).

7.3.11.7. All types of refinements of resolution given above are refutation complete. There exist also two forms of resolution restrictions which are incomplete. We mean here unit clause resolution and input clause resolution. The former was introduced by L. Wos, D. F. Carson and G. A. Robinson (1964). It permits resolution only when one parent is a unit clause, i.e. it consists of one literal. The input clause resolution was introduced by C. L. Chang (1970). It is a restricted form of linear resolution where one parent is always an input (given) clause. Chang proved that the unit clause and input clause resolutions are of equal power – they are complete over the class of Horn formulas.

7.3.11.8. Most mathematical theories contain among its symbols the equality symbol and among its axioms – the equality axioms. The immediate application of the usual resolution procedure generates in this case a lot of undesired clauses. Hence L. Wos and

G. A. Robinson (1970) introduced a procedure called paramodulation. This is the equality replacement rule with unification. It replaces all equality axioms except certain reflexivity axioms for functions. When paramodulation is restricted to replacement of the (usually) shorter term by the longer term with no instantiation allowed in the formula incorporating the replacement then one uses the term demodulation (cf. Wos-Robinson-Carson-Shalla 1967). It should be mentioned that there are also two other systems treating equality: system introduced by E. E. Sibert (1969) and E-resolution introduced by J. B. Morris (1969).

7.3.12. To finish this section we should add that almost simultaneosly with Robinson's invention of resolution, J. J͡u. Maslov in the USSR introduced a proof procedure very close to resolution in spirit. His method is called the inverse method and it is a test for validity rather than unsatisfiability (cf. Maslov 1964, 1971 and Kuehner 1971).

7.4. Development of mechanized deduction after 1965

The last section was devoted to the method of resolution and to its refinements. This approach to the mechanization and automatization of reasonings was dominating in the sixties. Nevertheless there were developed also other methods. In this section we shall discuss them briefly and sketch further development of the researches in the field after 1965.

7.4.1. We should begin with the Semi Automated Mathematics (SAM) project that spanned 1963 to 1967. It belongs to the human simulation approach (of. Section 7.1). In the framework of this project a succession of systems designed to interact with a mathematician was developed. They used many sorted ω-order logic with equality and λ-notation. The system SAM I was a proof checker but the theorem proving power continued to increase through SAM V which had substantial automatic capability. Only a part of the work in this project got recorded in the literature – cf. J. R. Guard-F. C. Oglesby-J. H. Bennett-L. G. Settle (1969).

7.4.2. Another project belonging to the human simulation approach developed in the mid-sixties was ADEPT, the Ph. D. thesis of L. M. Norton (cf. Norton 1966). It was a heuristic prover for group theory.

7.4.3. The dominant position of the resolution methods brought sharp criticism from some researchers. Their main argument was that there can not be a unique procedure which would suffice to realize (to simulate) the real intelligence. They stressed the necessity of using many components. One of those critics was M. Minsky from MIT. In 1970 C. Hewitt, a Ph. D. student at MIT wrote a dissertation on a new programmic language called PLANNER (cf. Hewitt 1971). Its goal was to structure a theorem prover system in such a way that locally distributed knowledge could be represented at various positions of the proving program. In fact it was not a theorem prover *per se*, but a language in which a "user" was to write his own theorem prover, specifically tailored to the problem domain at hand. This language was never fully implemented, only a subset of it was realized (microPLANNER).

7.4.4. About the same time another effort in human simulation approach was undertaken by A. Nevins (cf. Nevins 1974, 1975, 1975a). He built in fact at least two provers that were able to prove theorems which most resolution provers could not touch. For example Nevins could prove fully automatically that:

$$x^3 = e \longrightarrow f(f(a,b),b) = e.$$

where $f(x,y) = xyx^{-1}y^{-1}$. This result is much harder than the implication: "$x^2 = e \longrightarrow$ the group is abelian" which constituted the limit of capability of ADEPT.

7.4.5. We should mention also works of the group of scientists gathered around W. W. Bledsoe at the University of Texas which proved to be very important. They were working not for single uniform rule of inference for the whole mathematics but were seeking specific methods for particular domains of mathematics such as analysis, set theory or nonstandard analysis (cf. Bledsoe 1983, 1984).

7.4.6. So far we spoke about studies in seventies of the automated theorem proving which could be classified as human simulation approach. It does not mean that this was the only direction. There were also some new ideas within the logic approach. We should say here first of D. E. Knuth and P. B. Bendix (1970). They used the idea of rewrite rules – a device familiar to logicians. It is a replacement rule and allow to replace the left hand side by the right hand side at any occurrence of the left hand side. In equational theories one converts equations just to rewrite rules. Those rules enable us to reduce terms and to equate them. Knuth and Bendix proposed an algorithm which for a class of equational theories gave a complete set of rewrite rules – i.e. a set of rewrite rules sufficient to check the truth of every equation of the theory by demanding that equal terms reduce to the same normal form. An example of a theory to which the algorithm applies is the theory of groups (while the theory of abelian groups is not). There is an open problem connected with this algorithm: what equational theories have complete sets of rewrite rules?

7.4.7. Recall that the unification algorithm (described in the previous section) permits computation of a most general substitution for variables to make atomic formulas identical. Working with special theories, usually equational theories, one can simplify the procedure given by unification and resolution by introducing special unification. Its main idea is that the usual unification is augmented by equations or rewrite rules obtained from axioms of the considered theory. This idea was first formalized by G. D. Plotkin (1972) and further developed by M. E. Stickel (1981), (1985) and M. Livesey and J. Siekmann (1976). The idea of theory resolution of Stickel can be summerized as follows: since the resolution rule enlarges the whole number of steps, it is desirable to find macrorules in which certain sequences of steps could be performed in one step.

7.4.8. Another example of results which should be classified as logic oriented is the system of R. Overbeek developed later by S. Winker, E. Lusk, B. Smith and L. Wos and named AURA (Automated Reasoning Assistant). It was based on the old (i.e. coming from the sixties) ideas of unit preference, set-of-support for resolution, paramodulation and demodulation to which hyperresolution as well as more flexibility in demodulation and preprocessors for preparation of input from a variety of formats were added.

7.4.9. Around 1973 there appeared a very interesting effort different from the resolution approach and the strongly human oriented prover of Bledsoe. We mean here the Computational Logic Theorem Prover of R. S. Boyer and J. S. Moore (1975), (1979), (1981). This system uses the language of quantifier-free first order logic with equality and includes a general induction principle among the inference rules. It can be used to work within traditional mathematics (e.g. number theory) as well as to prove properties of programs and algorithms (so called proofs of correctness).

7.4.10. We should tell also about graph representation and about prover for systems of higher order. The former is based on the idea of enriching the structure of basic data with additional information, e.g. by representing the potential resolution steps in the graph structure. Literals or clauses and possible complementary literals form vertices of graphs which are connected by edges. This approach was introduced by R. Kowalski (1975) (cf. also S. Sickel 1976, R. E. Shostak 1976 and P. Andrews 1976).

7.4.11. The first proving system for higher order logics was developed by a group of scientists working under the direction of J. R. Guard in the early sixties. The studies were continued by W. E. Gould (1966), G. P. Huet (1975) and D. C. Jensen and T. Pietrzykowski (1976). The most important and influential group of people working in this direction is today at the Carnegie-Melon University (its chief is P. B. Andrews). They developed in the late seventies a theorem prover for type theory (TPS) (cf. Andrews 1981 and Miller, Cohen, Andrews 1982). It can prove for example Cantor's theorem as well as numerous first order theorems.

7.4.12. We shall finish this survey of activities in the field of mechanization and automatization of reasonings in sixties and seventies by mentioning an ambitous theorem prover now being developed at the University of Karlsruhe and named Markgraf Karl Refutation Procedure (cf. K. Bläsius, N. Eininger, J. Siekmann, G. Smolka, A. Herold, C. Walther 1981).

7.5. Final remarks

In the previous sections we have presented the recent period of the history of efforts to find an automated theorem prover. They were stimulated by the appearence of computers which made possible the practical realization of earlier ideas. The emphasis was put on the resolution procedure and the unification algorithm and their modifications because they proved to be the most influential ideas.

As was proved by Turing, Gödel and Church there exists no universal automated theorem prover for the whole mathematics – and even more, there are no such provers for most mathematical theories. Proving theorems in mathematics and logic is too complex a task for total automation because it requires insight, deep thought and much knowledge and experience. Nevertheless the semidecidability of mathematical theories was a sufficient motivation for looking for weaker theorem provers. We have described those efforts in the previous sections.

What does one expect from an automated theorem prover? First of all one obtains a certain unification of reasonings and their automatization. Having that one can shift the burden of proof finding from a mathematician and a logician to the computer. In this way we are also assured that faulty proofs would never occur. Are such automated theorem provers clever than people? Of course they can proceed quicker than a human being. But can they discover new mathematical results? The answer is YES. Some open questions have been answered in this way within finitely axiomatizable theories. For example S. Winker, L. Wos and E. Lusk (1981) answered positively the following open question: does there exist a finite semigroup which simultaneously admits of a nontrivial antiautomorphism without admitting a nontrivial involution? The progress in more complex theories such as analysis or set theory is slower, but there are also provers being able to prove some nontrivial theorems such as for example Cantor's theorem stating that a set has more subsets than elements (cf. P. Andrews, D. A. Miller, E. L. Cohen, F. Pfenning 1984) and various theorems in introductory analysis. The latter includes limit theorems of calculus such as

- the sum, product and composition of two continuous functions is continuous,
- differentiable functions are continuous,

- a uniformly continuous function is continuous,

as well as theorems of intermediate analysis (on the real numbers) such as

- Bolzano-Weierstrass theorem,
- if the function f is continuous on the compact set S then f is uniformly continuous on S,
- if f is continuous on the compact set S then $f[S]$ is compact,
- intermediate value theorem

(cf. Bledsoe 1984).

All those achievements can be treated as partial realizations and fulfilments of Leibniz's dreams of *characteristica universalis* and *calculus ratiocinator*. They are still far from what Leibniz did expect but they prove that a certain progress in the mechanization and automatization of reasonings and generally human thought has been made.

REFERENCES

Aarsleff, H. (1993). Descartes and Augustine on Genesis, Language, and the Angels. In: Dascal, M. and Yakira (Eds.), *Leibniz and Adam*. Tel Aviv: University Publishing Projects.

Ackermann, W. (1928). Über die Erfüllbarkeit gewisser Zahlenausdrücke. *Mathematische Annalen*, 100, 638-649.

Ajdukiewicz, K. (1934). Sprache und Sinn. *Erkenntnis*, IV, 100-138.

Anderson, R. and W. W. Bledsoe (1970). A Linear Format for Resolution with Merging and a New Technique for Establishing Completeness. *Journal of the Association for Computing Machines*, 17, 525-534.

Andrews, P. (1976). Refutations by Matings. *IEEE Trans. on Computers*, C–25, 801-807.

Andrews, P. (1981). Transforming Matings into Natural Deduction Proofs. *Proc. Fifth Conf. on Automated Deduction*, ed. by W. Bibel and R. Kowalski. *Lecture Notes in Computer Science 87*. Berlin - Heidelberg - New York: Springer-Verlag, pp. 281-292.

Andrews, P., D. A. Miller, E. L. Cohen and F. Pfenning (1984). Automating Higher-Order Logic. In: *Automated Theorem Proving. After 25 Years,* ed. by W. W. Bledsoe and D. W. Loveland, *Contemporary Mathematics*, 29, AMS, Providence, Rhode Island, pp. 169-192.

Aristotle (1949). *Aristotle's Prior and Posterior Analytics.* Revised text with Introduction and Commentary by D. W. Ross. Oxford: Clarendon Press. [As for the time of writing the original, the time interval 350-344 B.C. is regarded as most probable.]

Arndt, H. W. (1965). Einführung des Herausgebers [to] Christian Wolff, *Vernunftige Gedanken.* Hildesheim; Georg Olms [reprint].

Arnould, A. and P. Nicole (1662; 1965). *La logique ou l'art de penser contenant, outre les regles communes, plusieurs observations nouvelles, propres á former le jugement.* Edition critique présentée par Pierre Clair et François Girbal. Paris (1965): Presses Universitaires de France.

Automated Deduction, Lectures at 2nd International Summer School in Logic for Computer Science, July 4-15, 1994. Université de Savoie, France.

Bacon, F. (1620). *Novum Organum.* A modern critical edition by Th. Fowler in 1889, Clarendon Press, Oxford.

Basin, D. (1994). Induction Based on Rippling and Proof Planning. In: *Automated Deduction.*

Behmann, H. (1922). Beiträge zur Algebra der Logik, insbesondere zum Entscheidungsproblem. *Mathematische Annalen*, 86, 163-229.

Bernays, P. (1926). Axiomatische Untersuchungen des Aussagenkalküls der *Principia Mathematica. Math. Zeitschrift*, 25, 305-320.

Bernays, P. and M. Schönfinkel (1928). Zum Entscheidungsproblem der mathematischen Logik. *Mathematische Annalen*, 99, 342-372.

Bernoulli, Jacob and Johann Bernoulli (1685). *Parallelismus ratiocinii logici et algebraici.* Basileae.

Beth, E. W. (1955). *Semantic Entailment and Formal Derivability.* Amsterdam: Mededelingen der Koninklijke Nederlandse Akademie van wetenschapen, afd. letterkunde, new series, vol.18, no.13.

Beth, E. W. (1959). *The Foundations of Mathematics.* Amsterdam: North-Holland Publishing Company.

Bläsius, K., N. Eininger, J. Siekmann, G. Smolka, A. Herold and C. Walther (1981). The Markgraf Karl refutation procedure. In: *Proc. Seventh Intern. Joint Conf. on Artificial Intelligence*, pp. 511–518.

Bledsoe, W. W. (1983). Using Examples to Generate Instantiations for Set Variables. In: *Proc. Intern. Joint Conf. on Artificial Intelligence.*

Bledsoe, W. W. (1984). Some Automatic Proofs in Analysis. In: *Automated Theorem Proving. After 25 Years*, ed. by W. W. Bledsoe and D. W. Loveland, *Contemporary Mathematics*, 29, AMS, Providence, Rhode Island, pp. 89-118.

Bocheński, I. M. (1956). *Formale Logik.* Freiburg/ München: Alber.

Bonner, A. (Ed.) (1985). *Selected Works of Ramon Llull.* Princeton: University Press.

Boole, G. (1847). *The Mathematical Analysis of Logic, being an Essay toward a Calculus of Deductive Reasoning.* Cambridge: Macmillan, Barclay, and Macmillan and London: George Bell.

Boole, G. (1848). On the Calculus of Logic. *Cambridge and Dublin Mathematical Journal*, 3, 183-198.

Boole, G. (1854). *An Investigation of the Laws of Thought, on which are founded the Mathematical Theories of Logic and Probabilities.* London: Walton & Maberly.

Boole, M. E. (1905). Letters to a Reformer's Children [More bibliographical data in: (Peckhaus 1994)].

Boole, G. (1952). *Studies in Logic and Probability.* Ed. by R. Rhees. London: Watts & Co.

Boyer, R. S. (1971). *A Restriction of Resolution.* Ph. D. Thesis, University of Texas at Austin, Texas.

Boyer, R. S. and J. S. Moore (1975). Proving Theorems about LISP Functions. *Journal of the Association for Computing Machines*, 22, 129-144.

Boyer, R. S. and J. S. Moore (1979). *A Computational Logic.* New York: Academic Press.

Boyer, R. S. and J. S. Moore (1981). A Verification Condition Generator for FORTRAN. In: *The Correctness Problem in Computer Science*, ed. by R .S. Boyer and J. S. Moore. London.

Breger, H. (1988). Das Postulat der Explizierbarkeit in der Debatte um die künstliche Intelligenz. *Leibniz – Tradition und Aktualität. V. Internationaler Leibniz-Kongreß. Vorträge* Hannover: Leibniz Gesellschaft, pp. 107-116.

Brunus (Bruno), J. (G.) (1558). *De specierum scrutinio et lampade combinatoria Raymundi Lulli.* Pragae. In: (Zetznerus (Ed.) 1617).

Carrol, L. (1886). *The Game of Logic.* London and New York: Macmillan & Co.

Carrol, L. (1896). *Symbolic Logic.* London and New York: Macmillan.

Chandor, A. et al. (1985). *Dictionary of Computers.* London: The Penguin.

Chang, C. L. (1970). The Unit Proof and the Input Proof in Theorem Proving. *Journal of the Association for Computing Machines*, 17, 698-707.

Chang, C. L. and R. Lee (1973). *Symbolic Logic and Mechanical Theorem Proving.* New York - San Francisco - London: Academic Press.

Chinlund, T. J., M. Davis, G. Hineman and D. McIlroy (1964). *Theorem Proving by Matching.* Bell Laboratories.

Church, A. (1936). A Note on the Entscheidungsproblem. *Journal of Symbolic Logic*, 1, 40-41, 101-102.

Chwistek, L. (1921). Antynomje logiki formalnej [Antinomies of Formal Logic]. *Przegląd Filozoficzny*, 24, 164-171.

Chwistek, L. (1922). Zasady czystej teorii typów [The Principles of the Pure Theory of Types]. *Przegląd Filozoficzny*, 25, 354-391.

Chwistek, L. (1924-25). The Theory of Constructive Types. *Roczniki Polskiego Towarzystwa Matematycznego*, 2, 1924, 9-48; 3, 1925, 92-141.

Couturat, L. (1901). *La logique de Leibniz.* Paris: F. Alcan.

Couturat, L. (1903). *Opuscules et fragments inédits de Leibniz.* Paris. Reprinted 1961, Hildesheim: Olms.

Curry, H. B. (1952). A Theory of Formal Deducibility. *Journal of Symbolic Logic*, 17, 249-265.

Davies, P. (1993). *Princip Chaos. Die Neue Ordnung des Kosmos.* München: Goldmann. The English original – *Cosmic Blueprint*, London: Heinemann.

Davis, M. (1963). Eliminating the Irrelevant from Mechanical Proofs. *Proc. Symp. Applied Mathematics*, 25, 15-30.

Davis, M. (1988a). Mathematical Logic and the Origin of Modern Computers. In: Herken (Ed.), pp. 149-174.

Davis, M. (1988b). Influences of Mathematical Logic on Computer Science. In: Herken (Ed.), pp. 315-325.

Davis, M. and H. Putnam (1960). A Computing Procedure for Quantification Theory. *Journal of the Association for Computing Machines*, 7, 201-215.

Dedekind, R. (1888). *Was sind und was sollen die Zahlen?* Braunschweig: Friedrich Vieweg und Sohn.

De Morgan, A. (1847). *Formal Logic, or the Calculus of Inference, Necessary and Probable.* London: Taylor and Walton.

De Morgan, A. (1856). On the Symbols of Logic, the Theory of Syllogism, and in particular of the Copula, and the Application of the Theory of Probabilities to some Questions of Evidence. *Transactions of the Cambridge Philosophical Society*, 9, 79-127.

De Morgan, A. (1860). *Syllabus of a Proposed System of Logic.* London: Walton and Maberly.

De Morgan, A. (1861). Logic. In *English Cyclopedia* vol. 5, cols. 340-354.

De Morgan, A. (1864). On the Syllogism, no III, and on Logic in general. *Transactions of the Cambridge Philosophical Society*, 10, 173-230.

De Morgan, A. (1864). On the Syllogism, no IV, and on Logic in general. *Transactions of the Cambridge Philosphical Society*, 10, 331-358.

De Morgan, A. (1864). On the Syllogism no V, and on Logic in general. *Transactions of the Cambridge Philosophical Society*, 10, 428-488.

De Morgan, A. (1872). *Budget of Paradoxes.* Ed. by S. E. De Morgan. London: Longmans, Green, and Co.

Descartes, R. (1637). *Discours de la méthode.* Modern edition, by Ch. Adam and P. Tannery, in: *Oeuvres* vol. 6, Paris 1956.

Descartes, R. (1701). *Regulae ad directionem ingenii.* Amsterdam. Modern edition, by Ch. Adam and P. Tannery, in: *Oeuvres* vol. 10, Paris 1946.

De Solla Price, D. J. (1961). *Science since Babylon.* New Haven, London: Yale University Press.

Dipert, R. R. (1994). Leibniz's Discussion of Obscure, Confused, and Inadequate Ideas: its Relevance for Contemporary Cognitive Science. In: *Leibniz und Europa. VI. Internationaler Leibniz-Kongreß. Vorträge I Teil.* Hannover: Leibniz Gesellschaft, pp. 177-184.

Dreben, B. (1963). Corrections to Herbrand. *American Mathematical Society, Notices*, 10, 285.

Dreben, B. and P. Aanderaa (1964). Herbrand Analysing Functions. *Bulletin of the American Mathematical Society*, 70, 697-698.

Dreben, B., P. Andrews and P. Aanderaa (1963). Errors in Herbrand. *American Mathematical Society, Notices*, 10, 285.

Dreben, B., P. Andrews and S. Aanderaa (1963a). False Lemmas in Herbrand. *Bulletin of the American Mathematical Society*, 69, 699-706.

Dreben, B. and J. Denton (1966). A Supplement to Herbrand. *Journal of Symbolic Logic*, 31, 393-398.

Euler, L. (1768-1772). *Lettrès à une princesse d'Allemagne sur divers sujects de physique et de philosophie*, vol. 1, 2 and 3, St. Petersbourg: Academie impériale des sciences.

Frege, G. (1879). *Begriffsschrift, eine der arithmetischen nachgebildete Formelsprache des reinen Denkens.* Halle: Louis Nebert.

Frege, G. (1880/81). Booles rechnende Logik und die Begriffsschrift. [First published in: (Frege 1973)].

Frege, G. (1882). Über den Zweck der Begriffsschrift. *Sitzungsberichte der Jenaischen Gesellschaft für Medicin und Naturwissenschaft für das Jahr 1882*, pp.1-10.

Frege, G. (1884). *Die Grundlagen der Arithmetik, eine logisch-mathematische Undersuchung über den Begriff der Zahl.* Breslau: Wilhelm Koebner.

Frege, G. (1891). *Function und Begriff. Vortrag gehalten in der Sitzung vom 9. Januar 1891 der Jenaischen Gesellschaft für Medicin und*

Naturwissenschaft. Jena: H. Pohle. Reprinted in: Frege (1962), pp. 16-37.

Frege, G. (1892). Über Sinn und Bedeutung. *Zeitschrift für Philosophie und philosophische Kritik*, new series, 100, 25-50. Reprinted in: Frege (1962), pp. 38-63.

Frege, G. (1893). *Grundgesetze der Arithmetic, begriffsschriftlich abgeleitet*, I. Band. Jena: H. Pohle.

Frege, G. (1896). Über die Begriffsschrift des Herrn Peano und meine eigene. *Berichte über die Verhandlungen der Königlich Sächsischen Gesellschaft der Wissenschaften zu Leipzig, Mathematisch-philosophische Klasse*, 48, pp. 361-378.

Frege, G. (1902). Letter to Russell. First published in Heijenoort (1967), pp. 126-128. Cf. also Frege (1976), pp. 212-215.

Frege, G. (1903). *Grundgesetze der Arithmetik, begriffsschriftlich abgeleitet*, II. Band. Jena: H. Pohle.

Frege, G. (1950). *The Foundations of Arithmetic. A Logico-Mathematical Enquiry into the Concept of Number*. English translation of Frege (1884) with the German text, by J. L. Austin. Oxford: Blackwell.

Frege, G. (1962). *Funktion, Begriff, Bedeutung. Fünf logische Studien*. Hrsg. Günther Patzig. Göttingen: Vandenhoeck and Ruprecht.

Frege, G. (1973). *Schriften zur Logik. Aus dem Nachlaß*. Mit einer Einleitung von Lothar Kreiser. Berlin: Akademie Verlag.

Frege, G. (1976). *Nachgelassene Schriften und wissenschaftliche Briefwechsel*. Hrsg. H. Hermes, F. Kambartel, F. Kaulbach. *II. Band: Wissenschaftliche Briefwechsel*. Hamburg: Felix Meiner.

Gardner, M. (1958). *Logic, Machines and Diagrams*. New York – Toronto – London: McGraw-Hill.

Gelernter, H. (1959). Realization of a Geometry Theorem-Proving Machine. *Proc. Intern. Conf. on Information Processing*, Paris: UNESCO House. pp. 273–282. Also in: *Computers and Thought*, ed. by Feigenbaum and Feldman. McGrow–Hill.

Gelernter, H., J. R. Hanson and D. W. Loveland (1960). Empirical Explorations of the Geometry-Theorem Proving Machine. *Proc. Western Joint Computer Conf.*, pp. 143–147. Also in: *Computers and Thought*, ed. by Feigenbaum and Feldman. McGrow–Hill.

Gentzen, G. (1932). Über die Existenz unabhängige Axiomensysteme zu unendlichen Satzsysteme. *Mathematische Annalen*, 107, 329-350.

Gentzen, G. (1935). Untersuchungen über das logische Schliessen. *Mathematische Zeitschrift*, 39, 176-210, 405-431.

Gentzen, G. (1936). Über die Widerspruchsfreiheit der reinen Zahlentheorie. *Mathematische Annalen*, 112, 493-565.

Gentzen, G. (1938). Neue Fassung des Widerspruchsfreiheitsbeweis für die reine Zahlentheorie. *Forschungen zur Logik und zur Grundlagen der exakten Wissenschaften*, new series, no.4, 19-44. Leipzig: Hirzel.

Gentzen, G. (1943). Beweisbarkeit und Unbeweisbarkeit von Anfangsfällen der transfiniten Induktion in der reinen Zahlentheorie. *Mathematische Annalen*, 119, 140-161.

Gilmore, P. C. (1959). A Program for the Production of Proofs for Theorems Derivable within the First Order Predicate Calculus from Axioms. *Proc. Intern. Conf. on Information Processing.* Paris: UNESCO House.

Gilmore, P. C. (1960). A Proof Method for Quantification Theory: its Justification and Realization. *IBM Journal Research and Devel.* 28-35.

Gould, W. E. (1966). A Matching Procedure for ω-Order Logic. *Air Force Cambridge Research Laboratories, Report 66-781-4.*

Gödel, K. (1930). Die Vollständigkeit der Axiome des logischen Funktionenkalküls. *Monatshefte für Mathematik und Physik*, 37, 349-360.

Gödel, K. (1931). Über formal unentscheidbare Sätze der *Principia Mathematica* und verwandter Systeme, I. *Monatshefte für Mathematik und Physik*, 38, 173-198.

Greenstein, C. H. (1978). *Dictionary of Logical Terms and Symbols.* New York etc.: Van Nostrand Reinhold Co.

Guard, J. R., F. C. Oglesby, J. H. Bennett and L. G. Settle (1969). Semiautomated mathematics. *Journal of the Association for Computing Machines*, 16, 49-62.

Hamilton, W. (1859-1866). *Lectures on Metaphysics and Logic.* Ed. by H. L. Mansel and J. Veitch, vol. 1-4. Edinburgh - London: William Blackwood and Sons.

Harel, D. (1984). Dynamic Logic. In: D. Gabbay and F. Guenthner (Eds.) *Handbook of Philosophical Logic*, vol. 2. Dordrecht: Reidel (Synthese Library).

Herbrand, J. (1930). Recherches sur la théorie de la démonstration. *Travaux de la Société des Sciences et des Lettres de Varsovie.* Cl. III, no.33, 128pp.

Herbrand, J. (1931). Sur le problème foundamental de la logique mathématique. *Sprawozdania Towarzystwa Naukowego Warszawskiego, Wydział III*, 24, 12-56.

Herbrand, J. (1931a). Sur la non-contradiction de l'arithmetique. *Journal für die reine und angewandte Mathematik*, 166, 1-8.

Herken, R., (Ed.) (1988). *The Universal Turing Machine. A Half-Century Survey.* Oxford: Oxford University Press.

Hewitt, C. (1971). *Description and theoretical analysis (using schemata) of PLANNER: a language for proving theorems and manipulating models in a robot*, Ph.D. Thesis, MIT.

Hilbert, D. (1899). *Grundlagen der Geometrie.* In: Festschrift zur Feier der Enthüllung des Gauss-Weber-Denkmals. Leipzig.

Hilbert, D. (1900). Mathematische Probleme. *Vortrag, gehalten auf dem internationalen Mathematiker-Kongress zu Paris 1900, Nachrichten von der Königlichen Gesellschaft der Wissenschaften zu Göttingen.* 253-297; also: *Archiv der Mathematik und Physik, 3rd series*, 1 (1901), 44-63, 213-237.

Hilbert, D. (1904). Über die Grundlagen der Logik und Arithmetik. *Verhandlungen des Dritten Internationalen Mathematiker-Kongresses in Heidelberg vom 8. bis 13. August 1904, Leipzig 1905*, 174-185.

Hilbert, D. (1925). Über das Unendliche. *Mathematische Annalen*, 95, 161-190.

Hilbert, D. (1928). Die Grundlagen der Mathematik. *Abhandlungen aus dem mathematischen Seminar der Hamburgischen Universität*, 6, 65-85.

Hilbert, D. and W. Ackermann (1928). *Grundzüge der theoretischen Logik.* Berlin: Springer.

Hilbert, D. and P. Bernays (1934). *Grundlagen der Mathematik.* I Band. Berlin: Springer.

Hilbert, D. and P. Bernays (1939). *Grundlagen der Mathematik.* II Band. Berlin: Springer.

Hillgarth, J. N. (1971). *Ramon Lull and Lullism in Fourteenth Century France.* Oxford: Clarendon Press.

Hintikka, J. (1955). Form and Content in Quantification Theory. *Acta Philosophica Fennica*, 8, 7-55.

Hintikka, J. (1969). *The Philosophy of Mathematics.* Oxford: Oxford University Press.

Hocking, W. S. (1909). Two Expansions of the Use of Graphs in Elementary Logic. *University of California Publications in Philosophy*, 2, 31.

Holland, G. J. (1764). *Abhandlung über die Mathematik, die allgemeine Zeichenkunst und die Verschiedenheit der Rechnungsarten*. Tübingen.

Huet, G. D. (1975). A Unification Algorithm for Typed λ-Calculus. *Theoretical Computer Science*. 27-57.

Jaśkowski, S. (1934). On the Rules of Supposition in Formal Logic. *Studia Logica*, 1, 5-32.

Jensen, D. C. and T. Pietrzykowski (1976). Mechanizing ω-Order Type Theory through Unification. *Theoretical Computer Science*. 123-171.

Jevons, W. S. (1864). *Pure Logic, or the Logic of Quality apart from Quantity with Remarks on Boole's System and the Relation of Logic and Mathematics*. London - New York: E. Stanford.

Jevons, W. S. (1869). *The Substitution of Similars, the True Principle of Reasoning, Derived from a Modification of Aristotle's Dictum*. London: Macmillan and Co.

Jevons, W. S. (1870). On the Mechanical Performance of Logical Inference. *Philosophical Transactions of the Royal Society*, 160, 497-518.

Jevons, W. S. (1870). *Elementary Lessons in Logic*. London: Macmillan and Co.

Jevons, W. S. (1874). *The Principles of Science*, vol. 1-2. London: Macmillan and Co.

Jevons, W. S. (1874). *The Principles of Science. A Treatise on Logic and Scientific Method*. 2 vols. London: Macmillan and Co.

Jevons, W. S. (1886). *Letters & Journals of W. Stanley Jevons*, ed. by his wife. London: Macmillan and Co.

Jourdain, Ph. E. B. (1912). The Development of the Theories of Mathematical Logic and the Principles of Mathematics. *The Quarterly Journal of Pure and Applied Mathematics*, 43, 219-314.

Juniewicz, M. (1987). Leibniz's Modal Calculus of Concepts. In: (Srzednicki (Ed.) 1987), pp. 36-51.

Kalmár, L. (1932). Ein Beitrag zum Entscheidungsproblem. *Acta litterarum ac scientiarum Regiae Universitatis Hungaricae Francisco-Josephinae, Sectio scientiarum mathematicarum*, 5, 222-236.

Kalmár, L. (1934). Über einen Löwenheimschen Satz *Acta litterarum ac scientiarum Regiae Universitatis Hungaricae Francisco-Josephinae, Sectio scientiarum mathematicarum*, 7, 112-121.

Kalmár, L. (1936). Zurückführung des Entscheidungsproblem auf den Fall von Formeln mit einer einzigen, binären, Functionsvariablen. *Compositio Mathematica*, 4, (1937), 137-144.

Kennedy, H. C. (1973). What Russell Learned from Peano. *Notre Dame Journal of Formal Logic*, 14, 367-372.

Ketonen, O. (1944). *Untersuchungen zum Prädikatenkalkül, Annales Academiae Scientiarum Fennicae, ser. AI.* vol. 23, Helsinki, 77pp.

Klaus, G. and M. Buhr, (Eds.) (1969). *Philosophisches Wörterbuch.* Leipzig: Bibliographisches Institut.

Kleene, S. C. (1967). *Mathematical Logic.* New York – London – Sydney: John Wiley & Sons, Inc.

Kluge, E.-H. W. (1977). Frege, Leibniz et alii. *Studia Leibniziana*, Band IX/2.

Kluge, E.-H. W. (1980). *The Metaphysics of Gottlob Frege. An Essay in Ontological Reconstruction.* The Hague: Nijhoff.

Kluge, E.-H. W. (1980a). Frege, Leibniz and the Notion of an Ideal Language. *Studia Leibniziana*, Band XII/1.

Kneale, W. (1948). Boole and the Revival of Logic. *Mind*, 57, 149-175.

Kneale, W. and M. Kneale (1962). *The Development of Logic.* Oxford: Clarendon Press.

Knuth, D. E. and P. B. Bendix (1970). Simple Word Problems in Universal Algebra. In: *Combinatorial Problems in Abstract Algebras*, ed. by Leech, Pergamon - New York, pp. 263-270.

Kondakov, N. I. (1978). *Wörterbuch der Logik.* Berlin (West): Verlag das Europäische Buch.

Kotarbiński, T. (1964). *Leçons sur l'histoire de la logique.* Paris: Presses Univ. de France. Transl. from (Kotarbiński 1957) by A. Posner.

Kowalski, R. (1975). A Proof Procedure Using Connection Graph. *Journal of the Association for Computing Machines*, 22.

Kowalski, R. and D. Kuehner (1971). Linear Resolution with Selection Function. *Artificial Intelligence*, 2, 227-260.

Kreisel, G. (1953-54). A Variant to Hilbert's Theory of the Foundations of Arithmetic. *The British Journal of the Philosophy of Science*, 4, 107-129.

Kripke, S. (1959). A Completeness Theorem in Modal Logic. *Journal of Symbolic Logic*, 24, 1-14.

Kripke, S. (1959a). The Problem of Entailment. *Journal of Symbolic Logic*, 24, 324.

Kripke, S. (1963). Semantical Analysis of Modal Logic I: Normal Modal Propositional Calculi. *Zeitschrift für mathematische Logik und Grundlagen der Mathematik*, 9, 67-96.

Kuehner, D. G. (1971). A Note on the Relation between Resolution and Maslov's Inverse Method. In: *Machine Intelligence*, 6, ed. by Meltzer and D. Michie. New York, pp. 73-76.

Lambert, J. H. (1764). *Neues Organon oder Gedanken über die Erforschung und Bezeichnung des Wahren.* Leipzig.

Lambert, J. H. (1771). *Anlage zur Architectonic.* Riga.

Lambert, J. H. (1782, 1788; two successive volumes). *Logische und philosophische Abhandlungen.* Berlin.

Leibniz, G. W. (A - for the edition of the [German] Academy [of Sciences]). *Sämtliche Schriften und Briefe.* Ed. by the Prussian Academy of Sciences, continued by the Academy of Sciences of DDR in collaboration with the Leibniz-Archiv in Hannover, etc. Darmstadt, later Leipzig, Berlin. Started in 1923.

Leibniz, G. W. (1646). *Dissertatio de arte combinatoria in qua Ex Arithmeticae fundamentis Complicationum ac Transpositionum Doctrina novis praeceptis extruitur, et usus ambarum per universum scientiarum orbem ostenditur; nova etiam Artis Meditandi seu Logicae Inventionis semina sparguntur. Prefixa est Synopsis totius Tractatus, et additamenti loco Demonstratio* EXISTENTIAE DEI *ad Mathematicam certitudinem exacta.* Lipsiae [there follows the name and address of the printer]. In: (Leibniz A), vol. 1 (of Philosophical Writings) 1663-1672. Berlin 1971: Akademie-Verlag.

Leibniz, G. W. (1672). *Accessio ad arithmeticam infinitorum.* In: (Leibniz A), vol. 1 (of Mathematical Writings) 1672-1676. Berlin 1976: Akademie Verlag. [Written in 1672.]

Leibniz, G. W. (1684). Meditationes de Cognitione, Veritate et Ideis. In: Gerhardt, E. J., *Die Philosophische Schriften von Gottfried Wilhelm Leibniz*, vol. 4. Berlin 1880: Weidmannsche Buchhandlung, pp. 422-426. An English version in: G. W. Leibniz, *Philosophical Essays*, ed. and transl. by R. Askiew and D. Garber. Indianapolis 1989: Hacket Publishing Company. [The original appeared in *Acta Eruditorum*, Leipzig 1684.]

Leibniz, G. W. (1686). *Generales inquisitiones de analysi notionum et veritatum.* In: [a modern critical edition] *Allgemeine Untersuchungen über die Analyse der Begriffe und Wahrheiten. Herausgegeben, übersetzt und mit einem Kommentar versehen von Franz Schupp. Lateinisch – Deutsch.* Hamburg 1982: Felix Meiner Verlag. [The treatise written in 1686, first published in (Couturat 1903).]

Leibniz, G. W. (1714). *Principes de la [...] philosophie ou monadologie [...].* Ed. by A. Robinet. Paris 1954: PUF.

Leibniz, G. W. (1718). Principes de la nature et la grâce fondés en raison. Ed. by A. Robinet. Paris 1954: PUF. The original appeared in *L'Europe Savante*, vol. 4, part 1, November 1718. An English version in Parkinson (Ed.).

Leibniz, G. W. (1765). *Nouveaux essais sur l'entendement humain.* Ed. by R. E. Raspe. Amsterdam et Leipzig. [Written in 1704.]

Lenzen, W. (1984). Leibniz und Boolesche Algebra. *Studia Leibniziana*, XXI/2, 187-203.

Lenzen, W. (1987). Leibniz's Calculus of Strict Implication. In: (Srzednicki (Ed.) 1987), pp. 1-35.

Leśniewski, S. (1992). *Collected Works*, two vols. [translated from German and Polish]. Ed. by S. J. Surma, J. T. Srzednicki, D. I. Barnett and V. F. Rickey. Dordrecht: Kluwer.

Livesey, M. and J. Siekmann (1976). Unification of A+C-terms (bags) and A+C+I-terms (Sets). *Universität Karlsruhe, Interner Bericht Nr.5/76*, Karlsruhe.

Longman (1987). *Dictionary of Contemporary English.* London: Longman Group.

Lorenz, K. (1961). *Arithmetik und Logik als Spiele. Dissertation.* Köln.

Lorenzen, P. (1960). Logik und Agon. *Atti del XII Congresso Internationale di Filosofia, vol. IV.* Firenze.

Loveland, D. W. (1968). Mechanical Theorem Proving by Model Elimination. *Journal of the Association for Computing Machines*, 15, 236-251.

Loveland, D. W. (1969). A Simplified Format for the Model Elimination Procedure. *Journal of the Association for Computing Machines*, 16, 349-363.

Loveland, D. W. (1970). A Linear Format for Resolution. In: *Proc. IRIA Symp. on Automatic Demonstration*, Versailles, France 1968, LNM 125, Berlin - New York: Springer - Verlag, pp. 147-162.

Loveland, D. W. (1972). A Unifying View of some Linear Herbrand Procedures. *Journal of the Association for Computing Machines*, 19, 366-384.

Loveland, D. W. (1978). *Automated Theorem Proving: A Logical Basis.* Amsterdam - New York - Oxford: North-Holland Publ. Comp.

Loveland, D. W. (1984). Automated Theorem-Proving: A Quarter-Century Review. In: *Automated Theorem Proving. After 25 Years.* Eds. W. W. Bledsoe and D. W. Loveland, *Contemporary Mathematics*, 29, AMS, Providence, Rhode Island, pp. 1-46.

Löwenheim, L. (1915). Über Möglichkeiten im Relativkalkül. *Mathematische Annalen*, 76, 447-470.

Löwenheim, L. (1940). Einkleidung der Mathematik in Schröderschen Relativkalkül. *Journal of Symbolic Logic*, 5, 1-15.

Luckham, D. (1970). Refinements in Resolution Theory. In: *Proc. IRIA Symp. on Automatic Demonstration, Versailles*, France 1968, LNM 125, Berlin - New York: Springer-Verlag, pp. 163-190.

Lullus, R. (1617). *Ars magna et ultima.* In: (Zetznerus (Ed.) 1617).

Lullus, R. (1617). *Ars Brevis.* In: (Zetznerus (Ed.) 1617), (Bonner (Ed.) 1985).

Lycan, W. G. (Ed.) (1990). *Mind and Cognition. A Reader.* Oxford: Blackwell.

Łukasiewicz, J. (1925). Démonstration de la compatibilité des axiomes de la théorie de la déduction. *Annales de la Société Polonaise de Mathématiques*, 3, 149.

Łukasiewicz, J. (1930). Untersuchungen über den Aussagenkalkül. [In Reports of the Warsaw Scientific Society. The joint communication with A. Tarski. Engl. transl. in (Tarski 1956)].

Łukasiewicz, J. (1951). *Aristotle's Syllogistic from the Standpoint of Modern Formal Logic.* Oxford: Clarendon Press. 2nd ed. 1957.

Łukasiewicz, J. (1961). O determinizmie. In: *Z zagadnień logiki i filozofii. Pisma wybrane*, ed. by J. Słupecki. Warszawa: Państwowe Wydawnictwo Naukowe, pp. 114-126.

Łukasiewicz, J. (1970). On determinism. In: *Selected Papers*, ed. by L. Borkowski. Amsterdam - London: North-Holland Publ. Comp. and Warszawa: Państwowe Wydawnictwo Naukowe, pp. 110-128.

Macfarlane, A. (1885). The Logical Spectrum. *The London, Edinburgh and Dublin Philosophical Magazine*, 19, 286.

Macfarlane, A. (1890). Adaptation of the Method of the Logical Spectrum to Boole's Problem. *Proceedings of the American Association for the Advancement of Science*, 39, 57.

Makowsky, J. A. (1988). Mental Images and the Architecture of Concepts. In: Herken (Ed.), pp. 453-465.

Marciszewski, W. (1984). The Principle of Comprehension as a Present-Day Contribution to Mathesis Universalis. *Philosophia Naturalis. Archiv für Naturphilosophie und die philosophischen Grenzgebiete der exakten Wissenschaften und Wissenschaftsgeschichte.* Band 21, Heft 2-4, 523-537.

Marciszewski, W. (1993). Concepts-Processing as a Procedure of Analog-Digital Conversion. To appear in: I. Max and W. Stelzner (Eds.), *Logik und Mathematik. Frege-Colloquium 1993.* Berlin, New York: Walter de Gruyter.

Marciszewski, W. (1994). *Logic from a Rhetorical Point of View.* Berlin, New York: Walter de Gruyter.

Marciszewski, W. (1994a). A Jaśkowski-Style System of Computer-Assisted Reasoning. In: J. Woleński (Ed.), Philosophical Logic in Poland. Dordrecht, etc.: Kluwer (Synthese Library), pp. 85-102.

Marković, M. (1957). *Formalizam u savremenoj logici.* Beograd. Kultura.

Marquand, A. (1881). A Logical Diagram for *n* Terms. *The London, Edinburgh and Dublin Philosophical Magazine*, 12, 266.

Maslov, S. Ju. (1964). An Inverse Method of Establishing Deducibility in Classical Predicate Calculus. *Dokl. Akad. Nauk SSR.* 17-20.

Maslov, S. Ju. (1971). Proof-Search Strategies for Methods of the Resolution Type. *Machine Intelligence*, 6, ed. by B. Meltzer and D. Michie, New York, pp.77-90.

Mays, W., and D. P. Henry (1958). Jevons and Logic. *Mind*, 62, 484-505.

McColl, H. (1878, 1879, 1880). The Calculus of Equivalent Statements. *Proceedings of the London Mathematical Society*, 9, 9-10, 177-186; 10, 16-28; 11, 113-121.

Meltzer, B. (1966). Theorem-Proving for Computers: Some Results on Resolution and Renaming. *Computer Journal*, 8, 341-343.

Miller, D. A., E. L. Cohen and P. B. Andrews (1982). A Look at TPS. In: D. W. Loveland (Ed.), *Proc. Sixth Conf. on Automated Deduction.* Lecture Notes in Computer Science 138. Berlin - Heidelberg - New York: Springer-Verlag, pp. 60-69.

Moore, J. S. (1994). The Mechanization of Induction in the Boyer-Moore Theorem Prover. In: *Automated Deduction.*

Morris, J. B. (1969). E-resolution: Extension of Resolution to Include the Equality. In: *Proc. Intern. Joint Conf. Artificial Intelligence*. Washington D.C., pp. 287-294.

Murawski, R. (1985). Giuseppe Peano – Pioneer and Promotor of Symbolic Logic. *Komunikaty i Rozprawy Instytutu Matematyki UAM*, Poznań.

Murawski, R. (1987). Giuseppe Peano a rozwój logiki symbolicznej [Giuseppe Peano and the Development of Symbolic Logic]. *Roczniki Polskiego Towarzystwa Matematycznego, Seria II: Wiadomości Matematyczne*, 27, 261-277.

Nevins, A. J. (1974). A Human Oriented Logic for Automatic Theorem Proving. *Journal of the Association for Computing Machines*, 21, 606-621.

Nevins, A. J. (1975). Plane Geometry Theorem Proving Using Forward Chaining. *Artificial Intelligence*, 6, 1-23.

Nevins, A. J. (1975a). A Relaxation Approach to Splitting in an Automatic Theorem Prover. *Artificial Intelligence*, 6, 25-39.

Newell, A., J. C. Shaw and H. A. Simon (1956). Empirical Explorations of the Logic Theory Machine: A Case Study in Heuristics. In: *Proc. Western Joint Computer Conf.*, 15, pp. 218-239. Also in: Feigenbaum and Feldman (Eds.), *Computers and Thought*. McGrow-Hill 1963.

Newlin, J. W. (1906). A New Logical Diagram. *Journal of Philosophy, Psychology and Scientific Methods*, 3, 539.

Norton, L. M. (1966). *ADEPT – a Heuristic Program for Proving Theorems of Group Theory*, Ph. D. Thesis, MIT.

Nuchelmans, G. (1983). *Judgment and Proposition. From Descartes to Kant*. Amsterdam, etc.; North-Holland.

Oberschelp, A. (1992). *Logik für Philosophen*. Mannheim, etc.: Wissenschaftsverlag.

Parkinson, G. H. R. (Ed.) (1973). *Gottfried Wilhelm Leibniz: Philosophical Writings*. London, etc.: Dent, etc.

Peano, G. (1888). *Calcolo geometrico secondo l'Ausdehneungslehre di H. Grassmann, preceduto dalle operationi della logica deduttiva*. Torino: Bocca.

Peano, G. (1889). *Arithmetica principia nova methodo exposita*. Torino: Bocca.

Peano, G. (1889). *I principii di geometria logicamente esposti*. Torino: Bocca.

Peano, G. (1891). Principii di logica matematica. *Rivista di matematica*, 1, 1-10.

Peano, G. (1891a). Formule di logica matematica. *Rivista di matematica*, 1, 24-31, 182-184.

Peano, G. (1891b). Sul concetto di numero. *Rivista di matematica*, 1, 87-102, 256-267.

Peckhaus, V. (1994). Leibniz als Identifikationsfigur der Britischen Logiker des 19. Jahrhunderts. In: *Leibniz und Europa. VI. Internationaler Leibniz-Kongreß. Vorträge I Teil.* Hannover: Leibniz Gesellschaft, pp. 589-596.

Peirce, Ch. S. (1885). On the Algebra of Logic. A Contribution to the Philosophy of Notation. *The American Journal of Mathematics*, 7, 180-202. Reprinted in *Collected Papers*, ed. by C. Hartshorne and P. Weiss, Cambridge (Mass.), vol. 3, pp. 210-249 [Contains a Note which has not been published earlier].

Peirce, Ch. S. (1933-1934). *Collected Papers.* Ed. by C. Hartshorne and P. Weiss, Cambridge (Mass.): Harvard University Press.

Penrose, R. (1988). On the Physics and Mathematics of Thought. In: Herken (Ed.), pp. 491-522.

Penrose, R. (1989). *The Emperor's New Mind. Concerning Computers, Minds, and the Laws of Physics.* Oxford, etc.: Oxford University Press.

Plotkin, G. D. (1972). Building-in Equational Theories. In: B. Meltzer and D. Michie (Eds.), *Machine Intelligence*, 7, New York, pp. 73-90.

Ploucquet, G. (1763). *Methodus calculandi in logicis.* Tubingae.

Ploucquet, G. (1766). *Sammlung der Schriften, welche den logischen Calcul ... betreffen.* Frankfurt - Leipzig.

Popper, K. (1967). Epistemology without a Knowing Subject. In: B. van Rootselaar and J. F. Staal (Eds.) *Logic Methodology and Philosophy of Science III.* Amsterdam: North-Holland.

Popper, K. (1982). *Unended Quest. An Intellectual Autobiography.* Glasgow: Fontana/Collins.

Post, E. (1921). Introduction to a General Theory of Elementary Propositions. *The American Journal of Mathematics*, 43, 163-185.

Prantl, C. (1870). *Geschichte der Logik im Abendlande.* Vols. 1-4. Leipzig 1855–1870.

Prawitz, D. (1960). An Improved Proof Procedure. *Theoria*, 26, 102-139.

Presburger, M. (1929). Über die Vollständigkeit eines gewisses Systems der Arithmetik ganzer Zahlen, in welchem die Addition als einzige Operation hervortritt. In: *Sprawozdanie z I Kongresu Matematyków Krajów Słowiańskich - Comptus Rendus, I Congres des Math. des Pays Slaves.* Warszawa, pp. 92-101, 395.

Quine, W. V. O. (1955). A Proof Procedure for Quantification Theory. *Journal of Symbolic Logic*, 20, 141-149.

Ramsey, F. P. (1925). The Foundations of Mathematics. *Proceedings of the London Mathematical Society*, series 2, 25, 338-384.

Reiter, R. (1971). Two Results on Ordering for Resolution with Merging and Linear Format. *Journal of the Association for Computing Machines*, 18, 630-646.

Risse, W. (1964). *Die Logik der Neuzeit. I Band: 1500–1640.* Stuttgart-Bad Cannstatt: Friedrich Frommann Verlag.

Robinson, J. A. (1965). A Machine-Oriented Logic Based on the Resolution Principle. *Journal of the Association for Computing Machines*, 12, 23-41.

Robinson, J. A. (1965a). Automated Deduction with Hyperresolution. *International Journal Comput. Math.*, 1, 227-234.

Russell, B. (1900). *A Critical Exposition of the Philosophy of Leibniz.* Cambridge: The University Press.

Russell, B. (1902). Letter to Frege – first published in Heijenoort (1967), pp. 124-125.

Russell, B. (1903). *The Principles of Mathematics.* Cambridge: The University Press.

Russell, B. (1907). On Some Difficulties in the Theory of Transfinite Numbers and Other Types. *Proceedings of the London Mathematical Society*, series 2, 4, 29-53.

Russell, B. (1908). Mathematical Logic as Based on the Theory of Types. *American Journal of Mathematics*, 30, 222-262.

Russell, B. (1919). *Introduction to Mathematical Philosophy.* London-New York: George Allen and Unwin Ltd.

Russell, B. (1967). *The Autobiography of Bertrand Russell: 1872 – 1914.* London: George Allen and Unwin Ltd.

Ryle, G. (1949). *The Concept of Mind.* New York: Barnes & Noble.

Schnelle, H. (1988). Turing Naturalized: Von Neumann's Unfinished Project. In: Herken (Ed.), pp. 539-559.

Schoolem, G. G. (1955). *Major Trends in Jewish Mysticism*. London.

Schröder, E. (1880). Review of Frege (1879). *Zeitschrift für Mathematik und Physik*, 25, Historisch-literarische Abteilung. pp. 81-94.

Schröder, E. (1890, 1891, 1895, 1905). *Vorlesungen über die Algebra der Logik (Exakte Logik)*. Leipzig, vol. I 1890, vol. II/1 1891, vol. III 1895, vol. II/2 1905.

Schütte, K. (1951). Beweistheoretische Erfassung der unendlichen Induktion in der Zahlentheorie. *Mathematische Annalen*, 122, 369-389.

Schütte, K. (1956). Ein System des verknüpfenden Schliessens. *Archiv für mathematische Logik und Grundlagenforschungen*, 2, 56-67.

Shannon, C. E. (1938). A Symbolic Analysis of Relay and Switching Circuits. *Trans. AIEE* **57**, 713 ff.

Shea, W. R. (Ed.) (1983). *Nature Mathematized* [Proceedings of the 3rd International Congress of History and Philosophy of Science (Montreal 1980)]. Dordrecht: Reidel.

Shostak, R. E. (1976). Refutation Graphs. *Artificial Intelligence*, 7.

Sibert, E. E. (1969). A Machine Oriented Logic Incorporating the Equality Relation. In: B. Meltzer and D. Michie (Eds.), *Machine Intelligence*, 4, New York, pp. 103-134.

Sickel, S. (1976). Interconnectivity Graphs. *IEEE Trans. on Computers, C-25.*

Skolem, T. (1919). Untersuchungen über die Axiome des Klassenkalküls und über Produktations- und Summationsprobleme, welche gewisse Klassen von Aussagen betreffen. *Videnskapsselskapets skrifter, I. Mathematik-naturvidenskabelig klasse*, no.3.

Skolem, T. (1920). Logisch-kombinatorische Untersuchungen über die Erfüllbarkeit oder Beweisbarkeit mathematische Sätze nebst einem Theoreme über dichte Mengen. *Videnskapsselskapets skrifter, I. Mathematik-naturvidenskabelig klasse*, no.4.

Skolem, T. (1922). Einige Bemerkungen zur axiomatischen Begründung der Mengenlehre. *Matematikerkongressen i Helsingfors den 4 - 7 Juli 1922, Den Femte Skandinaviska matematikerkongressen, Redogörelse (Akademiska Bokhandeln, Helsinki 1923)*, pp. 217 - 232.

Skolem, T. (1923). Begründung der elementaren Arithmetik durch die rekurrirende Denkweise ohne Anwendung scheinbarer Veränderlichen mit unendlichen Ausdehnungsbereich. *Videnskapsselskapets skrifter, I. Matematik-naturvidenskabelig klasse*, no.6.

Skolem, T. (1928). Über die mathematische Logik. *Norsk matematisk tidskrift*, 10, 125-142.

Skolem, T. (1929). Über einige Grundlagenfragen der Mathematik. *Skrifter utgitt av Det Norske Videnskaps-Akademi i Oslo, I. Matematik-naturvidenskapelig klasse*, no.4.

Slagle, J. R. (1967). Automated Theorem Proving with Renamable and Semantic Resolution. *Journal of the Association for Computing Machines*, 14, 687-697.

Smith, G. C. (1982). *The Boole - De Morgan Correspondence 1842-1864*. Oxford: Clarendon Press.

Smullyan, R. (1968). *First-Order Logic*. Berlin: Springer-Verlag.

Spinoza, B. (1677). *Tractatus de intellectus emendatione*. The English translation by R. H. M. Elwes is found in the popular edition by D. D. Runes *How to Improve Your Mind*, 1956, New York: Philosophical Library.

Srzednicki, J. (Ed.) (1987). *Initiatives in Logic*. Dordrecht: Kluwer.

Stamm, E. (1935). *Z historii matematyki XVII wieku w Polsce* [From the history of mathematics of the 17th century in Poland]. Warszawa.

Stickel, M. E. (1981). A Complete Unification Algorithm for Associative-Commutative Functions. In: *Proc. Fourth Intern. Joint Conf. on Artificial Intelligence*, Tbilisi, USSR. Also in: *Journal of the Ass. for Computing Machines*, 28, 423-434.

Stickel, M. E. (1985). Automated Deduction by Theory Resolution. *Journal of Automated Reasoning*, 1, 333-356.

Swieżawski, S. (1974). *Dzieje filozofii europejskiej w XV wieku. Tom II: Wiedza* [The history of European Philosophy in the 15th century. Vol. II: Knowledge]. Warszawa: Akademia Teologii Katolickiej.

Śleszyński, J. (1925-1929). *Teorja dowodu. Podług wykładów uniwersyteckich prof. Jana Śleszyńskiego opracował S. K. Zaremba* [literally 'Proof Theory', in fact the book deals with Propositional Calculus], vol. 1 and 2. Kraków: Nakład Kółka Matematyczno-fizycznego U.J.

Tawney, R. N. (1942). *Religion and the Rise of Capitalism*. Pelican Books.

Tarski, A. (1930). Über einige fundamentale Begriffe der Metamathematik. [In Reports of the Warsaw Scientific Society. Engl. transl. in (Tarski 1956)].

Tarski, A. (1935). Der Wahrheitsbegriff in den formalisierten Sprachen. *Studia Philosophica*, vol. 1, pp. 261-405. Warszawa. [Engl. transl. in (Tarski 1956)].

Tarski, A. (1956). *Logic, Semantics, Metamathematics. Papers from 1923 to 1938.* Translated by J. H. Woodger. Oxford: Clarendon Press.

Tartaretus, P(etrus) (1517). *Commentarii in Isagogas Porphyrii et Libros Logicorum Aristotelis [...]*,. Basileae.

Turing, A. M. (1936-1937). On Computable Numbers, with an Application to the Entscheidungsproblem. *Proc. of the London Mathematical Society*, Series 2, vol. 42, pp. 230-265.

Turing, A. (1947). The Automatic Computing Engine. Lecture given to the London Mathematical Society on February 20, 1947. In: R. E. Carpenter and R. W. Doran (Eds.), *A. M. Turing's ACE Report and Other Papers.* Cambridge, MA: The MIT Press (1986).

Turski, W. M. (1985). *Informatics. A Propaedeutic View.* Warszawa: PWN /Amsterdam, etc.: North-Holland.

Van Evra, J. W. (1977). A Reassessment of George Boole's Theory of Logic. *Notre Dame Journal of Formal Logic*, 18, 363-377.

Van Heijenoort, J. (1967). *From Frege to Gödel. A Source Book in Mathematical Logic 1879-1931.* Cambridge, Massachusetts: Harvard University Press.

Van Peursen, C.-A. (1986). Ars Inveniendi bei Leibniz. *Studia Leibniziana*, Band XVIII/23e, pp. 183-194.

Venn, J. (1866). *The Logic of Chance. An Essay of the foundations and province of the theory of probability, with especial reference to its application to moral and social science.* London - Cambridge: Macmillan & Co.

Venn, J. (1880). On the Diagrammatic and Mechanical Representation of Propositions and Reasonings. *The London, Edinburgh and Dublin Philosophical Magazine and Journal of Science*, 5 Ser., 10, 1-18.

Venn, J. (1880). On the Forms of Logical Propositions. *Mind*, 5, 336.

Venn, J. (1881). *Symbolic Logic.* London: Macmillan & Co.

Von Neumann, J. (1951). The General and Logical Theory of Automata. In: Taub, A. H. (Ed.), *John von Neumann's Collected Works*, vol. 5. New York: Pergamon Press (1963).

Von Tschirnhaus, E. W. (1667). *Medicina mentis sive artis inveniendi praecepta generalia.* Amsterdam. 2nd ed. Leipzig 1695.

Wallen, L. (1994). Connection Methods and Theory of Proof Search. In: *Automated Deduction.*

Wang, H. (1960). Toward Mechanical Mathematics. *IBM Journal Research and Devel.*, 2-22. Also in: *Logic, Computers and Sets*, Chelsea, New York, 1970.

Wang, H. (1960a). Proving Theorems by Pattern Recognition. Part I. *Commun. Assoc. Comput. Mach.*, 3, 220-234.

Wang, H. (1961). Proving Theorems by Pattern Recognition. Part II, *Bell System Technical Jour.*, 40, 1-41.

Webster's Third New International Dictionary. Springfield, MA: Merriam Co. (1971).

Weingartner, P. (1983). The Ideal of Mathematization of All Sciences and of 'More Geometrico' in Descartes and Leibniz. In: (Shea (Ed.) 1983).

Whitehead, A. N. and B. Russell (1910). *Principia Mathematica.* Vol. I, Cambridge: Cambridge University Press.

Whitehead, A. N. and B. Russell (1912). *Principia Mathematica.* Vol. II, Cambridge: Cambridge University Press.

Whitehead, A. N. and B. Russell (1913). *Principia Mathematica.* Vol. III, Cambridge: Cambridge University Press.

Winker, S., L. Wos and E. Lusk (1981). Semigroups, Antiautomorphisms and Involutions: A Computer Solution to an Open Problem, I. *Math. of Computation*, 533-545.

Wittgenstein, L. (1922). *Tractatus Logico-Philosophicus.* London: Kegan Paul, Trench, Trubner & Co., Ltd.

Woleński, J. (1985). *Filozoficzna szkoła lwowsko-warszawska.* Warszawa: Państwowe Wydawnictwo Naukowe; English version: *Logic and Philosophy in the Lvov-Warsaw School.* Dordrecht: Kluwer Academic Publishers, 1989.

Wolff, Ch. (1712). *Vernünftige Gedanken von den Kraeften des menschlichen Verstandes und ihren richtigen Gebrauche in Erkenntnis der Wahrheit.* Halle.

Wos, L., D. F. Carson and G. A. Robinson (1964). The Unit Preference Strategy in Theorem Proving. *AFIPS Conf. Proc. 26*, Washington D.C., pp. 615–621.

Wos, L., G. A. Robinson and D. F. Carson (1965). Efficiency and Completeness of the Set of Support Strategy in Theorem Proving. *Journal of the Association for Computing Machines*, 12, 536-541.

Wos, L., G. A. Robinson, D. F. Carson and L. Shalla (1967). The Concept of Demodulation in Theorem Proving. *Journal of the Association for Computing Machines*, 14, 698-709.

Wos, L. and G. A. Robinson (1970). Paramodulation and Set-of-Support. In: *Proc. IRIA Symp. on Automatic Demonstration,* Versailles, France 1968, LNM 125, Berlin - New York: Springer-Verlag, pp. 276-310.

Yates, R., B. Raphael and T. Hart (1970). Resolution Graphs. *Artificial Intelligence*, 1, 257-289.

Zetznerus, L. (Ed.) (1617). *Raymundi Lulli Opera ea quae ad adinventam ab ipso Artem Universalem Scientiarum Artiumque omnium brevi compendio, firmaque memoria apprehendarum locupletissimaque vel oratione ex tempore pertractandarum, pertinent.* Argentorati.

INDEX OF SUBJECTS

INDEX OF NAMES

EXTENDED TABLE OF CONTENTS